Mammals of the Southwest

E. Lendell Cockrum

Drawings by Sandy Truett

The University of Arizona Press
Tucson, Arizona

About the Author . . .

E. Lendell Cockrum has spent more than 30 years in the field of biology and ecology. In 1952 he came to the University of Arizona from Kansas and became head of the University's Department of Ecology and Evolutionary Biology in 1977. As a researcher, teacher and consultant, his focus has been mainly on the desert and its inhabitants. Dr. Cockrum has filled numerous positions with the Smithsonian Institution, United Nations, and the U.S. government in international arid lands concerns. In addition, he has published more than a hundred articles and books on mammals, including *The Recent Mammals of Arizona: Their Taxonomy and Distribution*. He is a well-known specialist on bats.

Cover photograph by Francis Morgan, taken at the Arizona-Sonora Desert Museum, Tucson, Arizona.

THE UNIVERSITY OF ARIZONA PRESS

This book was set in 10/11 Pt. VIP Times Roman

Library of Congress Cataloging in Publication Data

Cockrum, E. Lendell.
Mammals of the Southwest.

Includes index. 1. Mammals—Southwest, New. I. Title.
QL719, S85C6 599.0979 81-21834
ISBN 0-8165-0760-0 AACR2
ISBN 0-8165-0759-7 (pbk.)

SANDY TRUETT
1943–1972

Mary C. Schlentz Truett grew up in Chicago and came to the University of Arizona as a biology student. Here she launched her career as a wildlife artist. As her work evolved, she became better known. A Houston gallery liked her work, displayed it and advertised it throughout the Southwest. In November, 1972, on her way home from a wildlife art show in Houston, Sandy was killed in an automobile accident.

Her goal as an artist was to create pictures that anyone could afford. She believed that regardless of their income or social standing, people were most important and that life was more interesting along less traveled routes. Referring to her life, she wrote, "Born a child in a land of roads, I learned to walk trails. . . ."

After seeing some of Sandy's pen and ink drawings, Dr. E. Lendell Cockrum decided he would like to collaborate on a book about Southwest mammals. Sandy produced the eighty drawings, and Dr. Cockrum furnished the words to accompany them. Her work justly deserves to be shared with others.

National Parks and Monuments

Contents

A Word From the Author

For years various people have expressed to me a need for a well illustrated guide to the mammals of Arizona and adjacent areas. This book is an attempt to fill their request. Its geographic scope includes much of the southwestern United States and northwestern Mexico. Most of the Sonoran and Chihuahuan deserts are included as well as adjacent desert grasslands, foothills and mountains.

The information should be equally useful to those interested in the mammals of west Texas, southeastern California, southern Utah or Colorado, southern Sonora or Chihuahua, or any area between. This is a large area encompassing more than 600,000 square miles.

The major patterns of mammalian development present in the Southwest are emphasized. Although at least 171 different species of native mammals have occurred in this region in historic times, only about eighty different major patterns are found. For example, most people have little difficulty recognizing the difference between a Round-tailed Squirrel and a Spotted Ground Squirrel—or distinguishing between a California Leaf-nosed Bat and a Brazilian Free-tailed Bat, even if they do not know their common names. These represent, then, what I term "major patterns." In contrast, even a professional mammalogist might have difficulty quickly distinguishing among a Cliff Chipmunk, Colorado Chipmunk, Gray-collared Chipmunk and others. Because they are morphologically similar—in size, shape, and color—only close attention to minor traits reveals that they are, in fact, distinct species. In this book single species of each "major type" are illustrated, and related species are briefly mentioned.

For each species illustrated, the common and scientific names, some major identifying features, measurements, habitats and highlights of life habits are given.

The common names and the scientific names used here conform with the 1979 *Revised Checklist of North American Mammals North of Mexico,* one of the Occasional Papers of The Museum at Texas Tech University.

Animals listed in the Contents are grouped into major biological orders—bats, rodents, hooved animals, etc—to aid in faster access and identification.

Measurements are given in English and metric system units. The values given are considered representative, not definitive. As in all animals, including man, individual variations make a single value almost meaningless. For instance, a young animal, a large, obese individual, or the sexual dimorphism that occurs in species such as the Wapiti (adult males are much larger than adult females)—all contribute to variability. However, if the values given are interpreted as being plus or minus ten percent, then most variation of adults will be included.

The distribution map shows the general region in which the mammal occurs or has occurred in the past. The notes on habitat indicate the special situations within this range where you would expect to find the animal.

A few native species that occur in the area are not mentioned even as a related species. One of these is man. Obviously the American Indians who live here are just as native as the other mammals listed. Also not included are the Pika *(Ochotona princeps)*, the Wolverine *(Gulo gulo)* or the Bison *(Bison bison)*. Drawings of these three were never made, since none occurs in Arizona. They only occur marginally in the area included in this book.

Also not listed are domesticated animals or the feral populations of domesticated animals such as the burro.

Various nonnative game animals have been introduced into parts of New Mexico, western Texas and Chihuahua. These include the Barbary Sheep *(Ammotragus lervia)* from North Africa, the Ibex *(Capra ibex)* from Eurasia, and Greater Kudu *(Tragelaphus capensis)* and the Oryx *(Oryx gazella)* also from Africa. None of these is included in this book.

Acknowledgments

A number of people have aided me in the preparation of this manuscript. I wish to thank especially Joe C. Truett for his permission to use Sandy's drawings. I also wish to thank Betty Bonanno, Marnetta Griffin and Eleanor Hoy for typing and proofreading the various versions of this manuscript and Elizabeth McCasland for transferring my sketch maps of species distribution to the final form.

My grateful appreciation is extended to the University of Arizona Press for publishing this book.

E. LENDELL COCKRUM

The Mammals

Collared Peccary

Dicotyles tajacu

Order Artiodactyla **Family Tayassuidae**

Identifying Features

A pig-like artiodactyl (hooved mammal) essentially tailless and with a well developed mane on shoulders and back. Males with well developed tusks. Hair stiff and a grizzled grayish to blackish often with a yellowish wash on hairs in the cheeks, throat and collar region.

Measurements

Total length, 37 inches (930 mm); tail, 1.8 inches (45 mm); hind foot, 7.1 inches (180 mm); ear, 3.5 inches (90 mm); weight, 60 pounds (27 kg).

Habitat

Generally in desert grasslands and desert areas, especially in areas of desert ranges and hills or areas of dense brush. Generally found near a free-flowing water source.

Life Habits

Usually gregarious, living in bands of three to twenty, sometimes more. Food is primarily a series of vegetable materials including roots, bulbs, shoots, and fruits. Cactus fruit, grain and even some animal material is eaten. These animals "bed down" in shallow holes in dense brush but also will use shallow caves and the entrances to abandoned mine tunnels as resting sites. One or two young are born in the spring or fall. At birth the young are reddish with a black dorsal stripe. Lives up to 15 years in captivity.

Collared Peccary

Wapiti or Elk

Cervus elaphus

Order Artiodactyla **Family Cervidae**

Identifying Features

Males of this large deer have antlers with at least five tines, of which the brow tine is well developed. Both males and females have a straw-colored tail. Males larger than the females. Females do not have antlers.

Measurements

Total length, 98 inches (2,500 mm); tail, 6 inches (150 mm); hind foot, 21 inches (550 mm); weight, usually about 600 pounds, but over 1,000 pounds known (270–450 kg).

Habitat

Higher mountain regions, often in high sheltered valleys.

Life Habits

Often in bands that migrate from high mountain meadows in the summer to lowlands in winter. Food is a variety of grasses, sedges and fresh growth of bushes and trees. A single calf, born in May or June, weighs about 30 pounds (13.5 kg) and is brownish with light spots. In captivity elk have lived as long as 22 years.

Related Species

Some consider that the large-antlered population of elk exterminated in Arizona and New Mexico by 1900 represented a distinct species, Merriam's Elk (*Cervus merriami*). Perhaps it was only subspecifically distinct. In any case, elk now in the region descended from introductions from the Yellowstone herd.

Wapiti or Elk

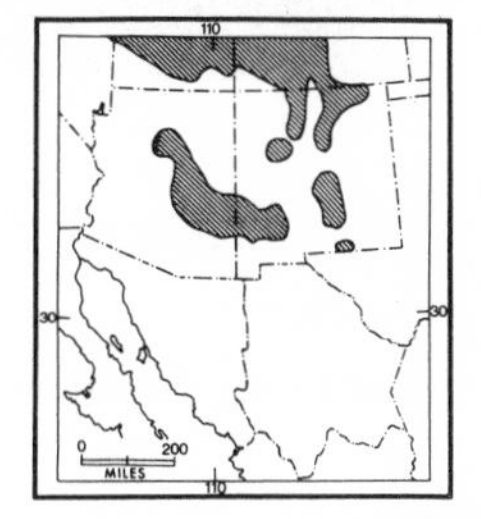

Mule Deer

Odocoileus hemionus

Order Artiodactyla **Family Cervidae**

Identifying Features

Similar to the Elk but smaller. The tail narrow and white with a black terminal tuft. Males have antlers with equally branched anterior and posterior tines. Young are light brown spotted with white.

Measurements

Total length, 63 inches (1,600 mm); tail, 7.9 inches (200 mm); hind foot, 17.7 inches (450 mm); ear, 7.9 inches (200 mm); weight, to 440 pounds (200 kg).

Habitat

Occur in broken country, brush and forest edge situations.

Life Habits

Food consists of a wide variety of browse especially from new growth on various woody plants. Some grazing is done on various grasses and forbs. One to three, usually two, fawns are born in May or June after a gestation period of about six months.

Mule Deer

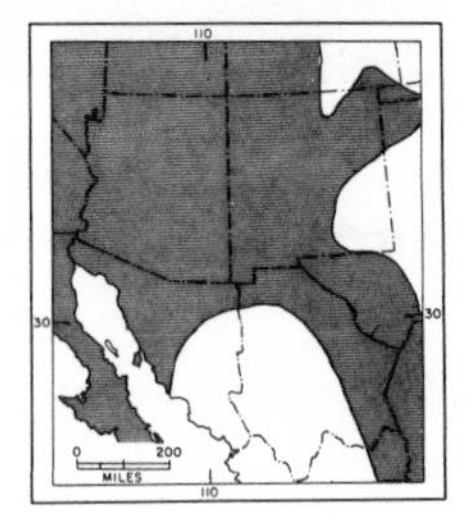

White-tailed Deer

Odocoileus virginianus

Order Artiodactyla **Family Cervidae**

Identifying Features

Similar to Mule Deer but with tail brown above, white beneath and white-tipped. Males with anterior beam of antlers long, thus equally divided antlers do not occur.

Measurements

Total length, 67 inches (1,700 mm); tail, 9.8 inches (250 mm); hind foot, 16 inches (410 mm); ear, 6.9 inches (175 mm); weight, to 325 pounds (150 kg).

Habitat

More commonly found in grasslands and level situations along forest edges or in riparian habitats.

Life Habits

Similar to the Mule Deer. Captive individuals live as long as 20 years.

White-tailed Deer

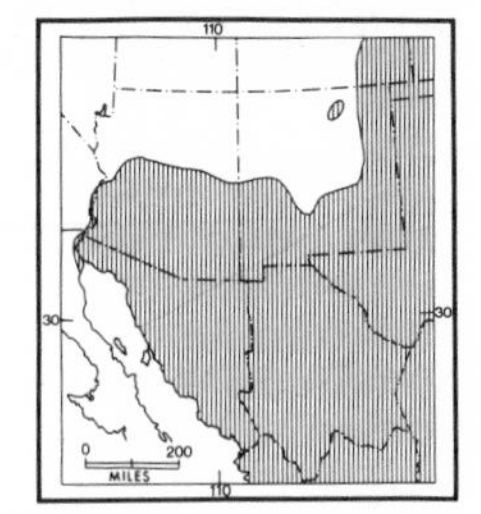

Pronghorn

Antilocapra americana

Order Artiodactyla **Family Antilocapridae**

Identifying Features

A medium-sized hooved mammal with unique branched horns in both males and females. The horns' outer sheath is shed annually. Color brownish with large rump patch that, when its hairs are erect, is almost white. Neck with horizontal white bands.

Measurements

Total length, 53 inches (1,350 mm); tail, 5.9 inches (150 mm); hind foot, 16.7 inches (425 mm); weight, to 125 pounds (57 kg).

Habitat

Generally shrub and grassland areas.

Life Habits

Food consists of a wide variety of browse including sagebrush and rabbitbrush. Some grasses are eaten. Often seen in herds, sometimes up to 60 in number. Have been clocked running 55 miles per hour (88 kph) for short distances. One or two young are born in June after a gestation period of about eight months. In captivity live up to 15 years. Now extinct in much of its former range.

Pronghorn

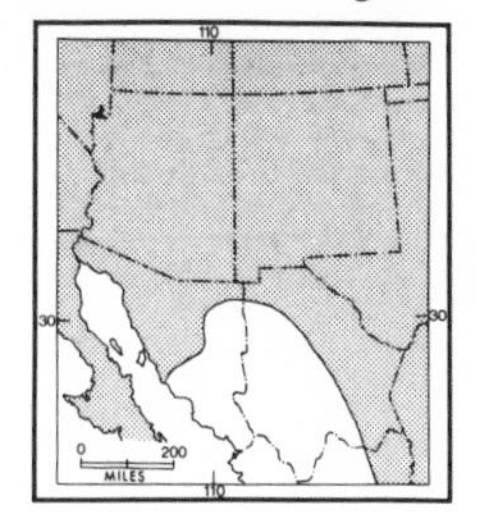

Mountain Sheep

Ovis canadensis

Order Artiodactyla **Family Bovidae**

Identifying Features

A large sheep with large, heavy curving horns in the rams. A characteristic white rump patch is present. Among many non-mammalogists this is generally called the Bighorn.

Measurements

Total length, 71 inches (1,800 mm); tail, 5.1 inches (130 mm); hind foot, 16.5 inches (420 mm); weight, 330 pounds (150 kg).

Habitat

Originally in foothills and mountains but, in competition with man, now generally restricted to rough mountainous areas.

Life Habits

Feeding occurs generally in the early morning. Food is a mixture of grasses and browse from a number of herbs and woody plants. Generally occurs in bands, often made up of 12 to 15 or even more individuals. They are always near a source of drinking water. After a gestation period of six months, one or two lambs are born in March or April. In captivity they live up to 15 years. Currently extinct in much of its former range.

Mountain Sheep

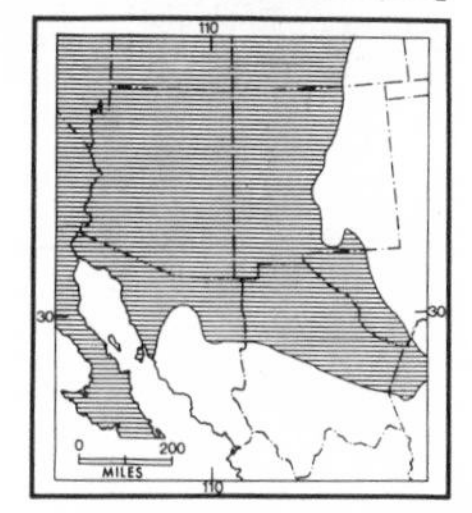

Coyote

Canis latrans

Order Carnivora **Family Canidae**

Identifying Features

This is a typical dog, closely resembling a small German Shepherd.

Measurements

Total length, 47 inches (1,200 mm); tail, 14 inches (350 mm); hind foot, 7.9 inches (200 mm); ear, 4.3 inches (110 mm); weight, 23 pounds (10.5 kg).

Habitat

Occurs in a wide range of habitats from high mountains to low deserts. Often numerous in the suburbs of cities, where they feed on garbage as well as other food.

Life Habits

Usually active in early morning and later afternoon. Some activity at night, especially on moonlit nights and on cooler overcast days. Food consists of a wide range of vegetable and animal matter. Juniper berries, cactus fruit, berries, rodents, rabbits, and insects are common. Some deer and domestic livestock are killed but mostly these are sick or have died from some other cause. A litter of up to 11 pups, generally five or six, is born in a burrow in the ground during the spring.

Coyote

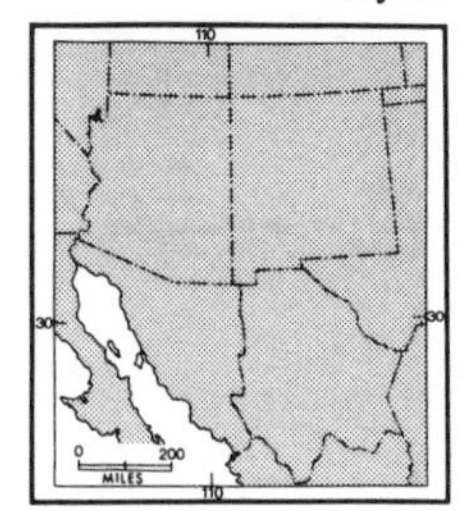

Gray Wolf

Canis lupus

Order Carnivora **Family Canidae**

Identifying Features
Similar to a large Coyote. Often similar in size and appearance to a big German Shepherd.

Measurements
Total length, 78 inches (1,980 mm); tail, 18 inches (450 mm); hind foot, 12 inches (300 mm); weight, 110 pounds (50 kg).

Habitat
Formerly occurred in most habitats.

Life Habits
Formerly occurred through most of this area. Probably now extinct. Scattered individuals were reported in southeastern Arizona and northern Mexico as late as the 1960s. Generally wolves live in small family groups of an adult female and male, young of the year and often the young of the previous litter. Group interactions, including cooperative hunting, are common. Formerly the annual hunting range of a wolf was as much as 50 miles (80 km) in length and perhaps as wide.

Related Species
Only the Red Wolf (*Canis rufus*), formerly in the southeastern United States, is considered to be a separate species. Timber Wolf, Lobo, and Mexican Wolf are all names applied to local populations of the Gray Wolf.

Gray Wolf

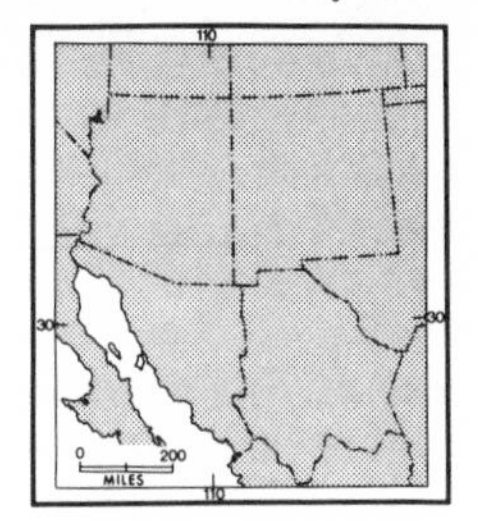

Red Fox

Vulpes vulpes

Order Carnivora **Family Canidae**

Identifying Features

A dog-shaped carnivore with long, pointed ears, an elongated muzzle and a very bushy tail. Dorsal color usually a reddish yellow shade. Feet black and tail white-tipped.

Measurements

Total length, 43 inches (1,100 mm); tail, 17 inches (430 mm); hind foot, 6.7 inches (170 mm); ear, 3.3 inches (85 mm); weight, 10 pounds (4.5 kg).

Habitat

Restricted to wooded areas in the northern part of the area.

Life Habits

This fox feeds mainly on small rodents and insects. Occasionally it captures a rabbit, and some fruits and berries are eaten. They generally live in burrows in the ground or in rock crevices. The same den is often occupied by generation after generation of foxes. A litter of one to eight pups is born in late spring or early summer. The male, female and young live and often hunt as a family unit until the young leave the following spring.

Red Fox

Kit Fox

Vulpes macrotis

Order Carnivora **Family Canidae**

Identifying Features

Like the Red Fox but smaller and with color lighter, the tail black-tipped and the feet not black.

Measurements

Total length, 31 inches (790 mm); tail, 9.8 inches (250 mm); hind foot, 4.9 inches (125 mm); ear, 3.3 inches (85 mm); weight, 3 pounds (1.5 kg).

Habitat

Desert and desert grasslands.

Life Habits

Similar to the Red Fox.

Related Species

The Swift Fox (*Vulpes velox*) occurs in the Great Plains grasslands.

1. Kit Fox

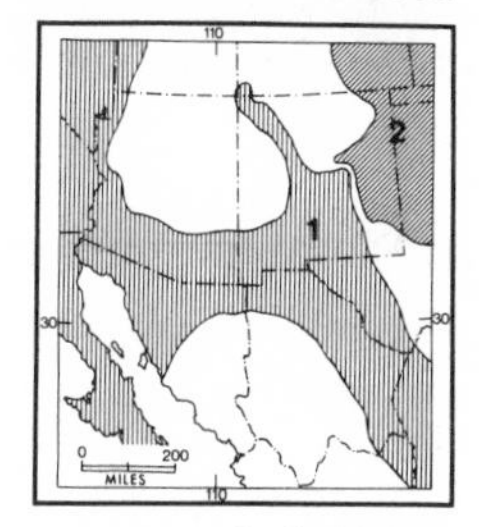

2. Swift Fox

Gray Fox

Urocyon cinereoargenteus

Order Carnivora **Family Canidae**

Identifying Features
Similar to Red Fox but tail with a dorsal black stripe made up of stiff black hairs. Dorsal color ranging from grizzled grayish to reddish.

Measurements
Total length, 37 inches (940 mm); tail, 16.7 inches (425 mm); hind foot, 5.3 inches (135 mm); ear, 2.6 inches (65 mm); weight, 8.4 pounds (3.8 kg).

Habitat
Occurs from 9,000 feet (2700 m) elevation down into low desert edge.

Life Habits
Similar to Red Fox.

Gray Fox

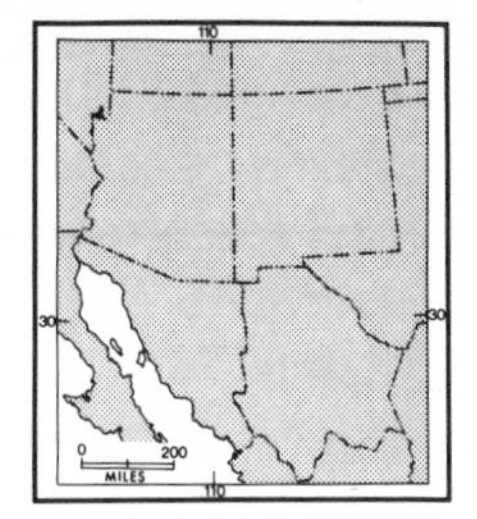

Black Bear

Ursus americanus

Order Carnivora **Family Ursidae**

Identifying Features

A large, short-tailed, heavily built carnivore, generally black in color.

Measurements

Total length, 63 inches (1,600 mm); tail, 3.5 inches (90 mm); hind foot, 9.5 inches (240 mm); ear, 4.9 inches (125 mm); weight, 500 pounds (225 kg).

Habitat

Now mostly restricted to remote, higher mountain regions. Formerly ranged down to desert edge, especially along streams.

Life Habits

Bears are omnivorous, feeding on a wide range of plant and animal material. Fruits, berries, acorns, various plant shoots, roots and bulbs are all eaten as are fish, insects, rodents and rabbits. Most larger animal food consists of carrion, either the remains of a mountain lion kill or an animal that died of other causes. Garbage is also readily taken. One to four cubs are born in the late winter, generally in a den in a hollow tree, rock crevice or shallow cave. Young are very small, blind and helpless at birth, weighing only about 1.5 pounds (0.7 kg). Sexual maturity is reached in the third year.

Black Bear

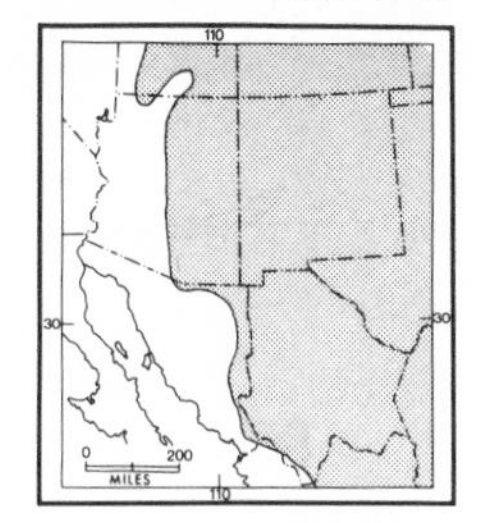

Grizzly Bear

Ursus arctos

Order Carnivora **Family Ursidae**

Identifying Features

Similar to the Black Bear but larger and color yellowish brown to black. Prominent hump at the shoulders.

Measurements

Total length, 85 inches (2,150 mm); tail, 4 inches (100 mm); hind foot, 12 inches (305 mm); ear, 5 inches (127 mm); weight, 900 pounds (400 kg).

Habitat

Formerly in mountains and along major streams in lowlands. Now extinct throughout this area.

Life Habits

Similar to the Black Bear. Known to live up to 34 years in captivity.

Sandy Truett

Ringtail

Bassariscus astutus

Order Carnivora **Family Procyonidae**

Identifying Features

A small-bodied, long-tailed somewhat cat-shaped carnivore with a bushy tail. The tail is strikingly marked with eight alternating white and light black rings and tipped with black.

Measurements

Total length, 31 inches (790 mm); tail, 15 inches (380 mm); hind foot, 3 inches (75 mm); ear, 2 inches (50 mm); weight, 2.2 pounds (1 kg).

Habitat

Generally in areas of rock exposure, most commonly in rocky canyons below 6,000 feet (1800 m) in elevation.

Life Habits

Food is primarily small rodents, especially mice and packrats, but includes various invertebrates. In season the fruit of *Opuntia* cactus as well as various berries are eaten. Two to four young are born in May or June, after a gestation period of about 53 days. Blind, helpless young are born in a nest in a rock crevice. Ringtails live up to eight years in captivity.

Ringtail

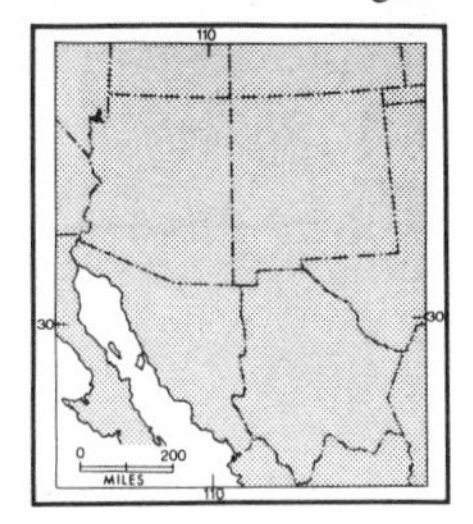

Raccoon

Procyon lotor

Order Carnivora **Family Procyonidae**

Identifying Features

A stout-bodied carnivore with a ringed, bushy tail somewhat like that of the Ringtail. Size much larger. Face marked with well developed black facial mask around the eyes. The tail has five to seven black rings and a black tip.

Measurements

Total length, 30 inches (750 mm); tail, 11 inches (280 mm); hind foot, 4.7 inches (120 mm); ear, 2.2 inches (55 mm); weight, 23 pounds (10.5 kg).

Habitat

Most common in areas where surface water is available. Often found along streams and in irrigated regions, generally at elevations below 6,000 feet (1800 m).

Life Habits

Activity is nocturnal, beginning usually after complete darkness. Feeding activities often take place along streams. Food includes various mice, birds, fish, invertebrates as well as grains, fruits, melons and other vegetable material. One to six, usually four, young are born after a gestation period of about two months. The young are blind and helpless at birth. Raccoons are known to live as long as 13 years in captivity.

Raccoon

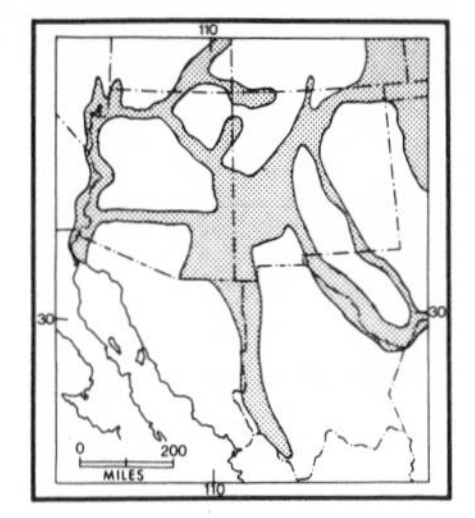

Coati

Nasua nasua

Order Carnivora **Family Procyonidae**

Identifying Features

Somewhat Raccoon-like but with a greatly elongated nose terminating in a flattened ''rooting'' pad; hind legs bigger than the forelegs and tail with rings barely present. Males are much larger than females.

Measurements

Total length, 41 inches (1,050 mm); tail, 20 inches (500 mm); hind foot, 3.6 inches (91 mm); ear, 1.6 inches (40 mm); weight, 25 pounds (11 kg).

Habitat

Oak woodlands and adjacent grassland edges in the southern part of this region.

Life Habits

Activity is generally diurnal. Often feed in bands of a few up to 50 individuals. They move across the forest floor, rooting in ground debris and exposing insects, small mammals, and various invertebrates. Coatis are excellent climbers and feed on fruits, berries, bird eggs and other foods found in trees. A litter of four to six young is born in the spring or early summer after a gestation period of about 75 days. The young are naked and blind at birth.

Coati

Long-tailed Weasel

Mustela frenata

Order Carnivora **Family Mustelidae**

Identifying Features

A small, short-legged, long-bodied carnivore with a black-tipped tail about half the length of the head and body. The dorsal color is brownish. Variable whitish patches are sometimes present on the face. Males are much larger than females.

Measurements

Total length, 16 inches (410 mm); tail, 5 inches (130 mm); hind foot, 5.7 inches (50 mm); ear, 0.8 inch (21 mm); weight, 9.5 ounces (270 g).

Habitat

A variety of habitats in the mountainous portions of the region. Absent from the desert grasslands and the deserts proper.

Life Habits

Activity is mainly at night but occasionally seen during the day. Food consists mainly of rodents, rabbits, and birds and bird eggs. Pocket gophers, captured as the weasel makes its way through the burrow systems, make up much of its food. Four to eight, usually five, young are born in a nest in the ground generally in mid-April.

Related Species

Three other members of this family occur in the area. All have restricted distribution. Mink (*Mustela vison*), generally found near streams, weigh up to three pounds (1.3 kg); Ermine (*Mustela erminea*), small, weighing only about two to three ounces (57–86 g); Marten (*Martes americana*), weighs up to 12 pounds (5.4 kg).

Long-tailed Weasel

110
30
30
0
200
MILES
110
Sandy Truett

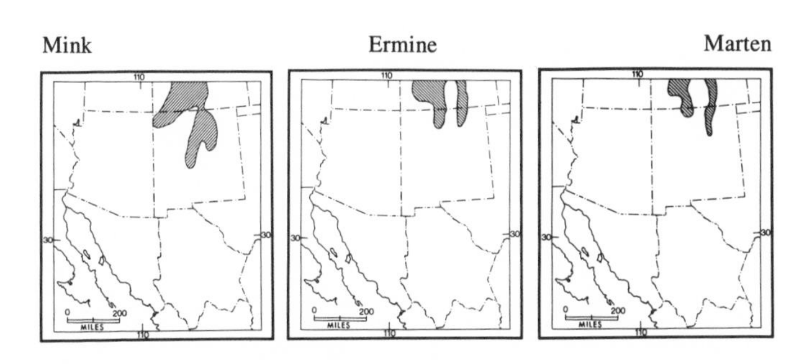
Mink
Ermine
Marten
110
30
30
0
200
MILES
110

Black-footed Ferret

Mustela nigripes

Order Carnivora **Family Mustelidae**

Identifying Features
A relatively large specialized weasel, over 1.5 feet (0.5 m) in length. Light in color, generally a pale yellowish buff blending to almost white on the throat and ventral surface. The face mask, feet and tip of the shortish tail are brownish black to black.

Measurements
Total length, 19.3 inches (490 mm); tail, 3.9 inches (100 mm); hind foot, 2.4 inches (60 mm); ear, 1.2 inches (31 mm); weight, 1.5 pounds (680 g).

Habitat
Generally in association with prairie dog towns. Now extinct through most of its former range.

Life Habits
The natural history of this rare animal has never been adequately documented. According to several authors, this species may depend on prairie dog burrows for shelter, and prairie dogs may be the major food source. It is now on the Rare and Endangered Species List of the United States Fish and Wildlife Service. Any sight observations should be reported to a game biologist or park naturalist.

Black-footed Ferret

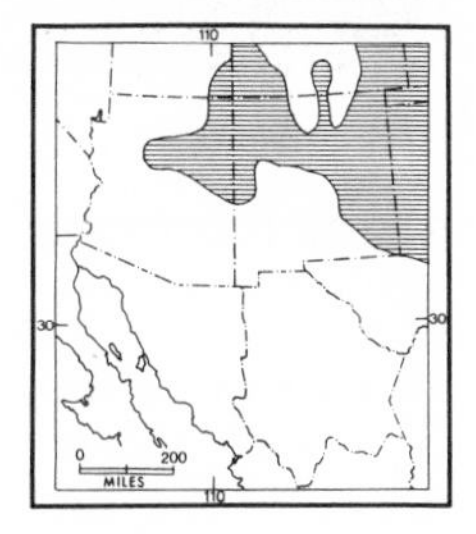

Badger

Taxidea taxus

Order Carnivora **Family Mustelidae**

Identifying Features

A stout-bodied carnivore well adapted for burrowing. Legs, neck, tail and ears are short. The forefeet have long, strong digging claws. The head is broad and triangular. The head is black with white cheeks and a white middorsal line, the back a grizzled mixture of black and white hairs.

Measurements

Total length, 31 inches (780 mm); tail, 5.1 inches (130 mm); hind foot, 4.3 inches (110 mm); ear, 2 inches (50 mm); weight, 20 pounds (9.1 kg).

Habitat

Occupies a range of non-forest habitats from high mountains to low deserts.

Life Habits

Activity is generally at dusk or at night, but sometimes badgers are active during the day. Food consists mainly of various small rodents, reptiles and insects that are dug from burrows in the ground. One to five, usually two, young are born in March or April. The young are blind at birth but are covered by fur. In captivity they are known to live for 13 years.

Badger

Western Spotted Skunk

Spilogale gracilis

Order Carnivora **Family Mustelidae**

Identifying Features
A small, slim-bodied skunk with white spots variously scattered in the black pelage. Tail usually tipped in white.

Measurements
Total length, 17 inches (430 mm); tail, 5.9 inches (150 mm); hind foot, 1.8 inches (45 mm); ear, 0.8 inch (20 mm); weight, 14 to 21 ounces (400 to 600 g).

Habitat
Generally in rough, broken country below 8,000 feet (2400 m) in elevation. In the desert mainly restricted to riparian situations.

Life Habits
Active generally at night. Food consists primarily of insects and other invertebrates. A litter of two to ten, usually four, young is born in March or April.

Western Spotted Skunk

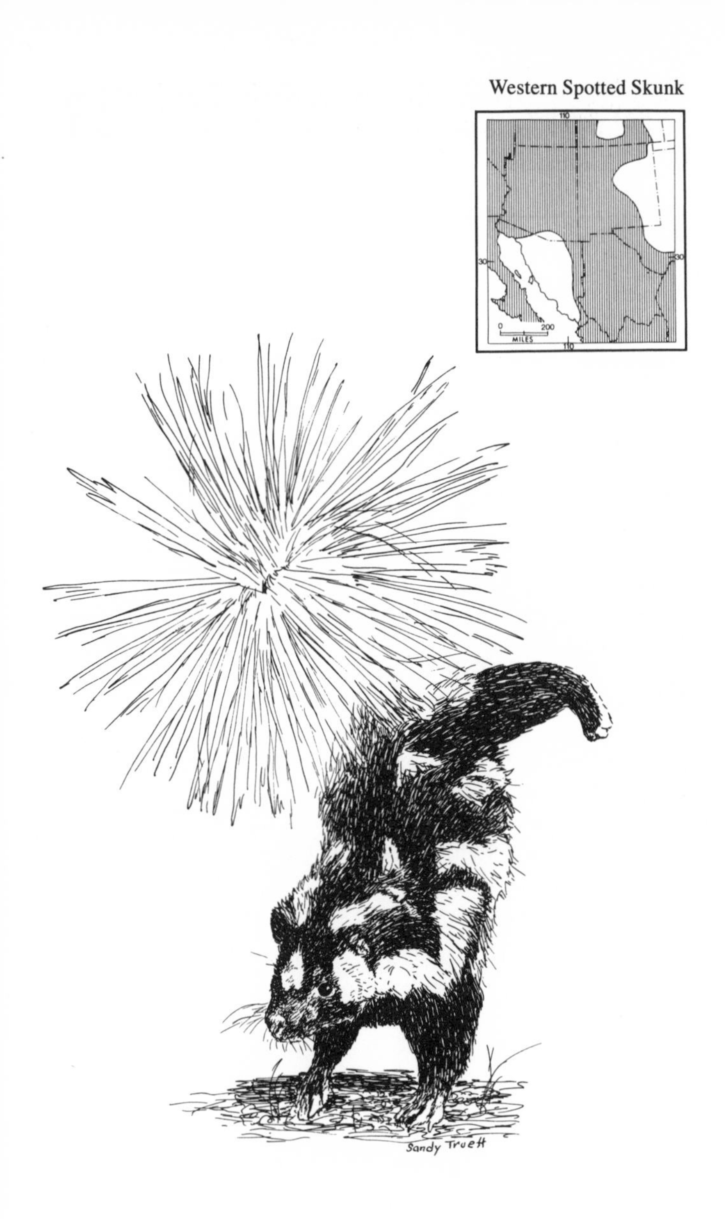

Striped Skunk

Mephitis mephitis

Order Carnivora **Family Mustelidae**

Identifying Features
A cat-sized carnivore, black with white on the head, neck, shoulders, back and part of the large bushy tail. The mid-back area is black. The legs are short.

Measurements
Total length, 28 inches (700 mm); tail, 11.4 inches (290 mm); hind foot, 3 inches (75 mm); ear, 1.1 inches (28 mm); weight, 6.6 pounds (3 kg).

Habitat
Occurs in a wide range of habitats from high mountains to low deserts.

Life Habits
Activity is generally at night. Food consists of a wide range of plant and animal foods. Carrion is readily taken as are various fruits, bird eggs, insects, worms, and other invertebrates. A litter of four to ten, generally five, young is born in May or June. A captive animal is known to have lived for at least six years.

Striped Skunk

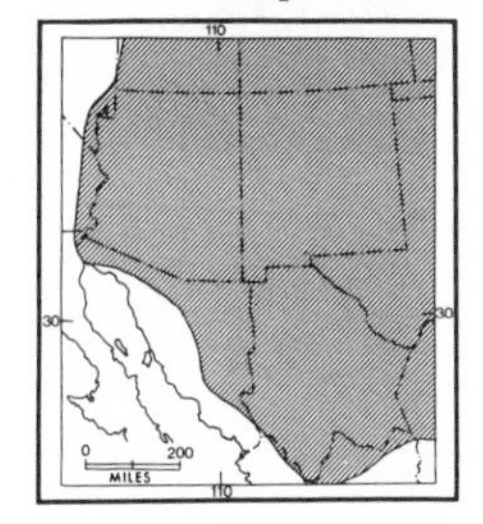

Hooded Skunk

Mephitis macroura

Order Carnivora **Family Mustelidae**

Identifying Features

Similar to Striped Skunk. Two color phases are known, a black phase with widely separated small white lines and a white phase that has white mixed with black over much of the back and tail. Smaller than the Striped Skunk.

Measurements

Total length, 26.8 inches (680 mm); tail, 12.2 inches (310 mm); hind foot, 2.6 inches (65 mm); ear, 1.2 inches (30 mm); weight, 2 pounds (900 g).

Habitat

Generally in more rocky situations than the Striped Skunk, in the south central part of this region.

Life Habits

Similar to the Striped Skunk. Diet probably of smaller items, including more insects and small rodents than that of the Striped Skunk. Grasshoppers are often taken.

Hooded Skunk

Hog-nosed Skunk

Conepatus mesoleucus

Order Carnivora **Family Mustelidae**

Identifying Features

Similar to the Striped Skunk but with nose pad well developed. White stripe on back generally much wider than in the Striped Skunk.

Measurements

Total length, 29 inches (734 mm); tail, 9.1 inches (232 mm); hind foot, 3 inches (77 mm); ear, 1 inch (25 mm); weight, 5 pounds (2.3 kg).

Habitat

Restricted to the mountain foothills and desert grasslands of the southern part of the region.

Life Habits

Similar to the Striped Skunk.

Hog-nosed Skunk

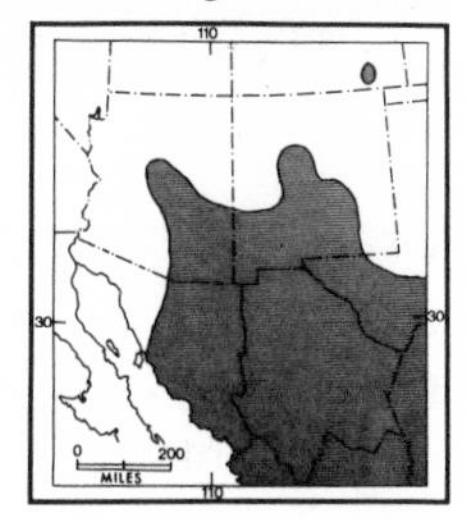

River Otter

Lutra canadensis

Order Carnivora **Family Mustelidae**

Identifying Features

A large carnivore adapted for a life in and around water. The body is somewhat teardrop in shape, ears small and can be closed. Tail thick at the base and attenuated, toes are webbed and can be spread broadly. Sole of foot has tufts of hair, especially under toes. The pelage is dense and waterproof. Color is brownish above and somewhat paler below.

Measurements

Total length, 51 inches (1,300 mm); tail, 32 inches (815 mm); hind foot, 5.7 inches (146 mm); weight, 20 pounds (9.1 kg).

Habitat

Restricted to areas of permanent water, generally along major streams. Now extinct in much of its former range.

Life Habits

Generally solitary and nocturnal. One individual may feed along 6 to 15 miles (7–24 km) of river in a single night. Dens are generally a hole in a river bank, often under the roots of a tree from which the soil has been partially eroded. Food is mainly fish and aquatic invertebrates but some frogs, birds, mammals and vegetable matter are taken. Two to five, usually two or three, young are born blind and naked, generally in later spring. Captive animals have lived as long as 15 years.

Related Species

The Southern River Otter (*Lutra annectens*) has been taken in the southern part of this region.

1. River Otter

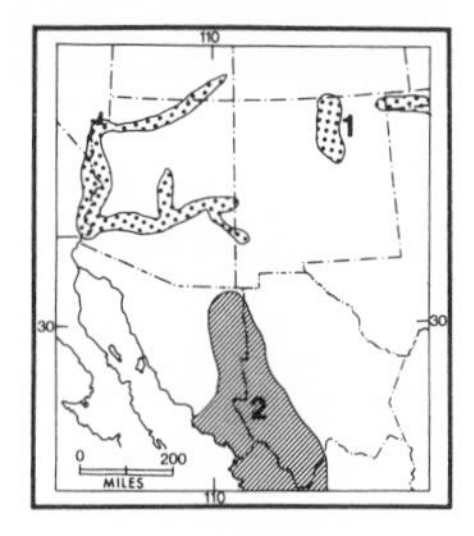

2. Southern River Otter

Jaguar

Felis onca

Order Carnivora **Family Felidae**

Identifying Features
A large spotted cat about the size of the Old World Leopard. Some black (melanistic) individuals have been recorded. Males larger than females.

Measurements
Total length, 90 inches (2,300 mm); tail, 26 inches (650 mm); hind foot, 10.8 inches (275 mm); weight, 285 pounds (130 kg).

Habitat
Formerly present in the desert grassland and lower mountains of the southwestern part of the region. Now rarely seen.

Life Habits
Generally a solitary animal that is active mainly at night. As is true of most large carnivores, these cats require a large hunting range, perhaps covering one to two hundred square miles (260–518 square kilometers) in a year. Food consists of various large animals, especially javelina, deer, and pronghorn. Some domestic animals are killed for food. Jaguars also feed along rivers and sometimes scoop fish out of the water. Two to four kittens are born in a litter after a gestation period of 100 days. In captivity they survive up to 22 years.

Jaguar

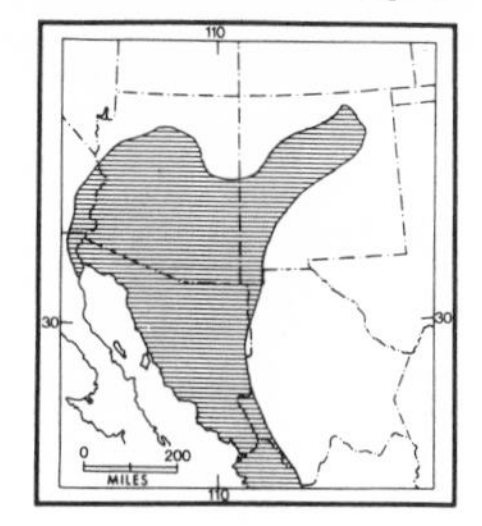

Mountain Lion

Felis concolor

Order Carnivora **Family Felidae**

Identifying Features
A large, unspotted, leopard-sized cat with short yellowish-brown fur.

Measurements
Total length, 90 inches (2,300 mm); tail, 31 inches (800 mm); hind foot, 12 inches (300 mm); ear, 3.9 inches (100 mm); weight, 160–220 pounds (73–100 kg).

Habitat
Formerly in most of the region. Now restricted to the more remote mountainous areas.

Life Habits
Similar to the Jaguar. Both large cats are often accused of killing domesticated animals including sheep, goats, and calves. Some are killed by these cats, but probably not nearly as many as often reported. Populations of large cats were never high, despite various "old-timer" tales. Now populations are so low that extinction is probable in much of the region.

Mountain Lion

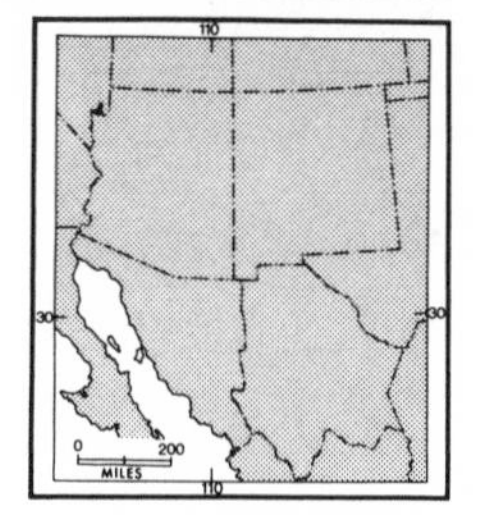

Ocelot

Felis pardalis

Order Carnivora **Family Felidae**

Identifying Features
A medium-sized, long-legged spotted cat with a ringed tail.

Measurements
Total length, 47 inches (1,200 mm); tail, 14.8 inches (375 mm); hind foot, 6.7 inches (170 mm); weight, to 35 pounds (16 kg).

Habitat
Generally restricted to densely wooded areas. Now probably extinct in this part of its former range.

Life Habits
Mainly nocturnal, the ocelot often hunts in pairs and feeds on various rodents, rabbits, birds, reptiles and frogs. Litters of two young are often born in a grass-lined nest in a hollow log or tree after a gestation period of about 90 days.

Ocelot

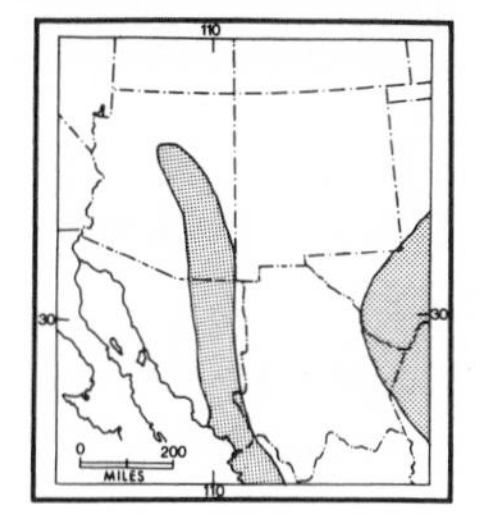

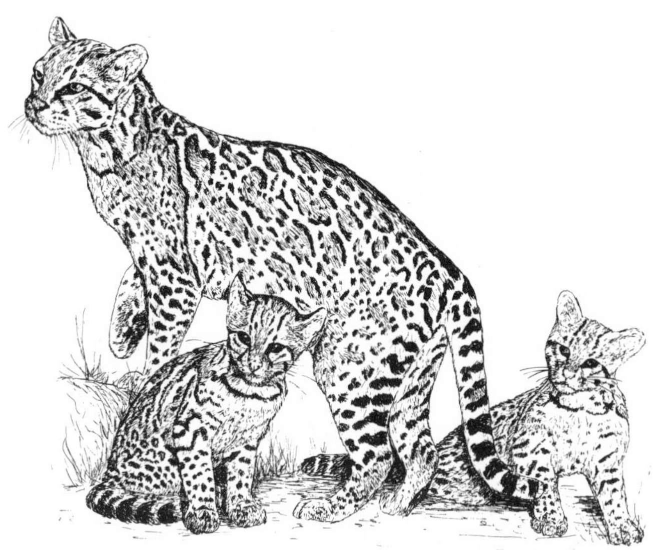

Jaguarundi

Felis yagouaroundi

Order Carnivora **Family Felidae**

Identifying Features
A medium-sized, short-legged, long-bodied, unspotted cat with a long tail. Two color phases occur, a reddish brown and an iron gray.

Measurements
Total length, 47 inches (1,200 mm); tail, 20 inches (500 mm); hind foot, 5.5 inches (140 mm); ear, 1.2 inches (30 mm); weight, 20 pounds (9 kg).

Habitat
Generally along streams in chaparral and forest areas. Now greatly reduced in numbers in this area.

Life Habits
Similar to the Ocelot.

Jaguarundi

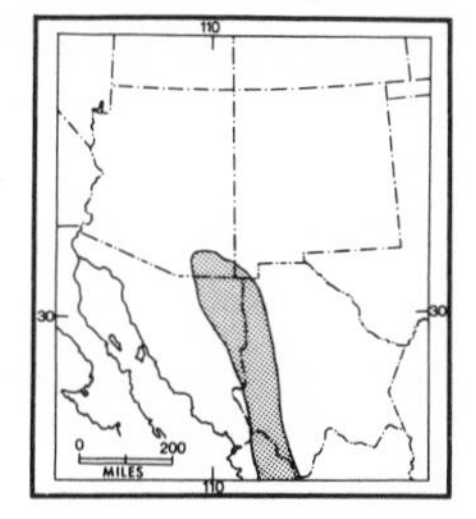

Bobcat

Felis rufus

Order Carnivora **Family Felidae**

Identifying Features
A short-tailed cat. Ears have tufts, paws are large, the head is broad and large. Color a reddish brown with black spots.

Measurements
Total length, 37 inches (940 mm); tail, 6 inches (150 mm); hind foot, 7.5 inches (190 mm); ear, 3.1 inches (80 mm); weight, 24 pounds (10.9 kg).

Habitat
Generally in brush and forest edge situations. Rocky canyons are often inhabited.

Life Habits
Activity is mainly nocturnal but sometimes active during the day. Food consists of various small rodents, rabbits, and occasionally lambs and young goats. Dens are often in rock crevices or in the ground. One to four kittens are born in a litter, usually in April or May.

Related Species
Lynx (*Felis lynx*), larger, generally in forests above 7,000 feet (2100 m).

Bobcat

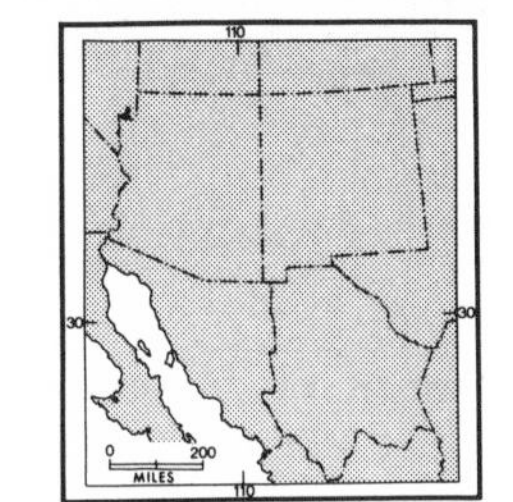

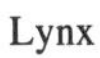

Lynx

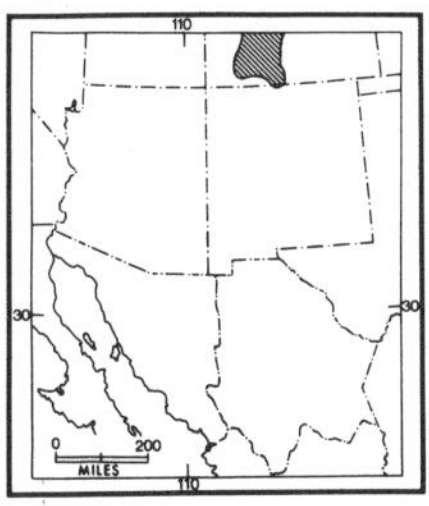

Least Chipmunk

Eutamias minimus

Order Rodentia **Family Sciuridae**

Identifying Features

These chipmunks have five evenly spaced longitudinal dark stripes on the back, one short lateral dark stripe on each side, and a long, well-haired tail. The ventral surface is usually buffy, bright orange or grayish yellow. The ears are long and pointed.

Measurements

Total length, 7.7 inches (195 mm); tail, 3.5 inches (90 mm); hind foot, 1.2 inches (30 mm); ear, 0.6 inch (16 mm); weight, 1.8 ounces (50 g).

Habitat

Generally lives at higher elevations in sagebrush areas, some distance from conifers.

Life Habits

Food consists of a wide variety of seeds, berries, cactus fruit, fungi and other vegetable material. In places juniper berries, acorns and pine nuts are used. At times some insects are also eaten. Four to six young per litter is common and some females have two litters in a year.

Related Species

The various kinds of chipmunks generally live at different elevations and in varying habitats. Some occur as high as the tree zone, others as low as the desert edge. The five related species in this area differ slightly in size and color. Cliff Chipmunk (*Eutamias dorsalis*); Colorado Chipmunk (*Eutamias quadrivittatus);* Gray-collared Chipmunk *(Eutamias cinereicollis)*; Gray-footed Chipmunk *(Eutamias canipes);* and Uinta Chipmunk *(Eutamias umbrinus)*.

Least Chipmunk

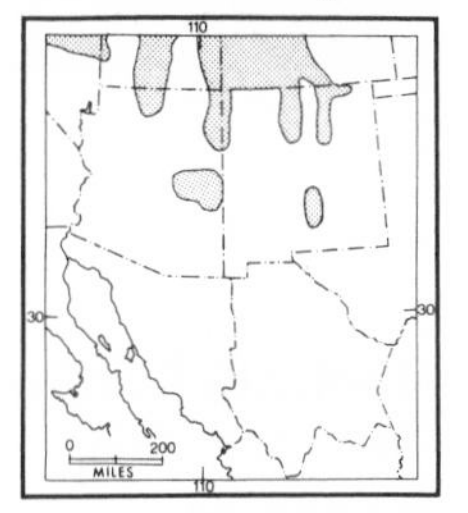

Cliff Chipmunk

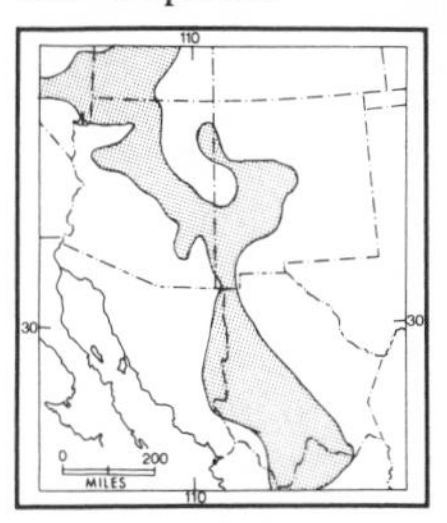

1. Uinta Chipmunk
2. Colorado Chipmunk
3. Gray-collared Chipmunk
4. Gray-footed Chipmunk

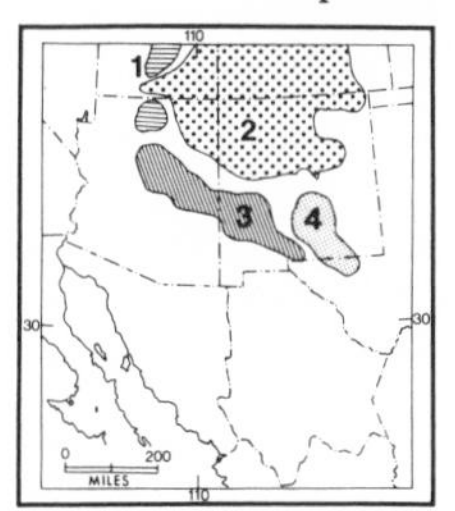

Harris' Antelope Squirrel

Ammospermophilus harrisii

Order Rodentia **Family Sciuridae**

Identifying Features

Differ from other small squirrel-like rodents in that they have a well developed white stripe extending on each side from the shoulder to the hip. The tail is usually held curved over the back, exposing the buffy colored under surface.

Measurements

Total length, 9.1 inches (230 mm); tail, 3 inches (75 mm); hind foot, 1.5 inches (38 mm); ear, 0.5 inch (12 mm); weight, 4.4 ounces (125 g).

Habitat

Restricted to desert-like areas in southern Arizona and northwestern Sonora.

Life Habits

Diurnal animals, active above ground most of the year. On very cold winter days they remain below ground; on hot summer days they are most active in the early morning. Food consists almost entirely of green vegetation. Buds and new growth of mesquite and various cactus fruit are seasonal favorites. Five to nine young are born in a litter, usually in February or March.

Related Species

Two related species are known in this area: Texas Antelope Squirrel (*Ammospermophilus interpres*) and White-tailed Antelope Squirrel (*Ammospermophilus leucurus*).

1. Harris' Antelope Squirrel

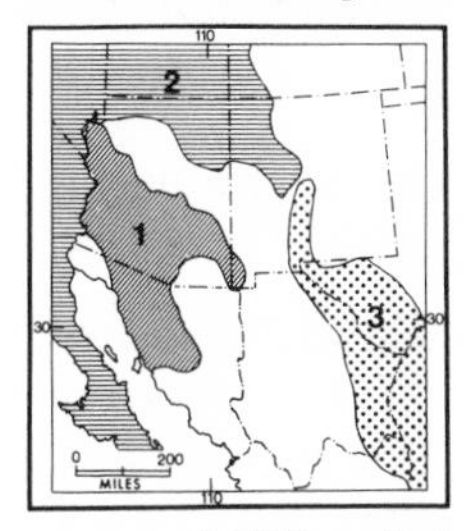

2. White-tailed Antelope Squirrel
3. Texas Antelope Squirrel

Thirteen-lined Ground Squirrel

Spermophilus tridecemlineatus

Order Rodentia **Family Sciuridae**

Identifying Features

This medium-small ground squirrel has a characteristic pattern of alternating dark and light stripes. A row of squarish white spots is in the center of each of the dark stripes. The dark stripes vary in color from brown to blackish. The tail is bushy.

Measurements

Total length, 9.3 inches (235 mm); tail, 2.9 inches (74 mm); hind foot, 1.3 inches (33 mm); ear, 0.4 inch (9 mm); weight, 4.6 ounces (130 g).

Habitat

Restricted to grassland areas, generally in well-drained sites with short grasses present.

Life Habits

These solitary animals live in burrows in grasslands. No mound exists at the mouth of the burrow. Food consists of seeds, green plant growth and many insects. Hibernation may last five to six months. Mating occurs in the spring and, after a gestation period of 28 days, a litter of five to thirteen young is born in an underground nest.

Thirteen-lined Ground Squirrel

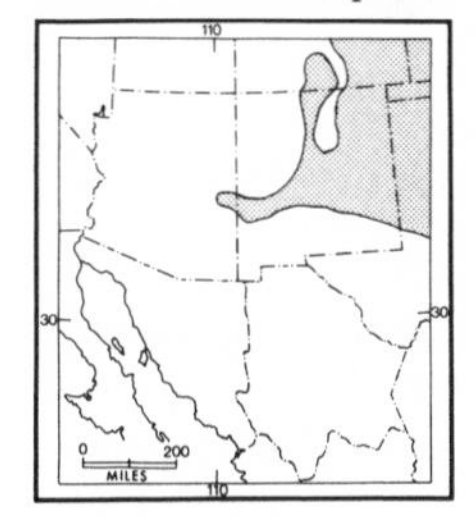

Spotted Ground Squirrel

Spermophilus spilosoma

Order Rodentia **Family Sciuridae**

Identifying Features

Similar to the Thirteen-lined Ground Squirrel in general size and shape. The coat color is a light brown, indistinctly spotted on the back and rump by a series of roundish white spots.

Measurements

Total length, 9.8 inches (250 mm); tail, 3.1 inches (80 mm); hind foot, 1.4 inches (35 mm); ear, 0.4 inch (11 mm); weight, 3.5 ounces (100 g).

Habitat

Generally associated with dry sandy soils in dry and desert grasslands.

Life Habits

Similar to the Thirteen-lined Ground Squirrel. Food consists of plant material, especially seeds, and insects. They live in burrows that usually have the entrances hidden under a bush. Generally only one lives in a burrow. Two litters of five to twelve each are born each year, one in the spring and one in the summer. Much of the winter is spent in hibernation.

Spotted Ground Squirrel

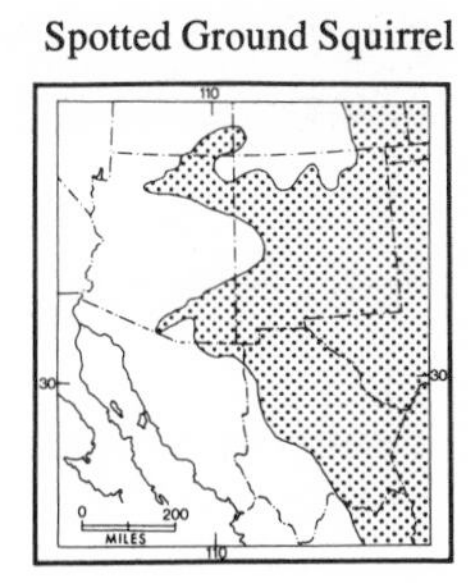

Rock Squirrel

Spermophilus variegatus

Order Rodentia **Family Sciuridae**

Identifying Features

This large ground squirrel, with a long bushy tail and mottled gray back, is generally confused with a tree squirrel. It differs, however, in having shorter ears and a tail less bushy and lacking a definite white fringe.

Measurements

Total length, 18.9 inches (480 mm); tail, 7.9 inches (200 mm); hind foot, 2.4 inches (61 mm); ear, 1.2 inches (30 mm); weight, 1 pound 11 ounces (700 g).

Habitat

Occurs generally in rocky situations from high mountains to the desert edge.

Life Habits

This large ground squirrel feeds on the flowers, fruits and seeds of various plants. Because they require rock exposures or steep walled arroyos for burrow construction, populations in a region tend to be concentrated in restricted areas even though these animals are not strictly colonial. At higher elevations they are inactive during the colder part of the year. Young are born during the summer, with young of the year first appearing above ground as early as late June and as late as mid-September.

Rock Squirrel

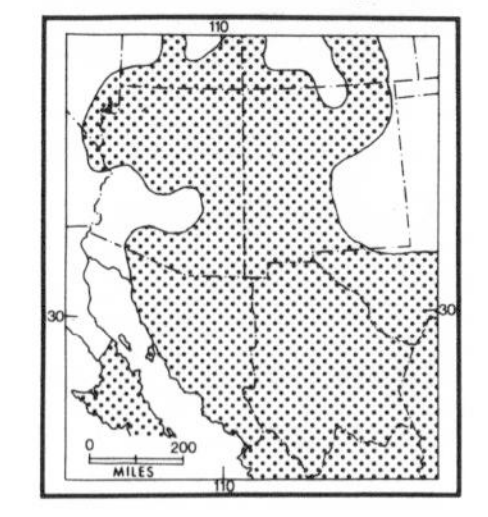

Round-tailed Ground Squirrel

Spermophilus tereticaudus

Order Rodentia **Family Sciuridae**

Identifying Features

This small ground squirrel is a uniform light color above, generally some shade of cinnamon. The tail is covered with short hairs and is round.

Measurements

Total length, 9.5 inches (240 mm); tail, 2.8 inches (70 mm); hind foot, 1.4 inches (35 mm); ear, 0.2 inch (6 mm); weight, 4.9 ounces (140 g).

Habitat

Generally found in sandy soils of the lower parts of the Sonoran Desert.

Life Habits

These desert dwelling ground squirrels feed primarily on green vegetation but also eat seeds, flowers and cactus fruit. One litter born in April is usual, but some females may have a second litter in July. Four to twelve young per litter have been recorded. Most of the winter months are spent in hibernation.

Related Species

One related species occurs in the general region: Mexican Ground Squirrel (*Spermophilus mexicanus*).

1. Round-tailed Ground Squirrel

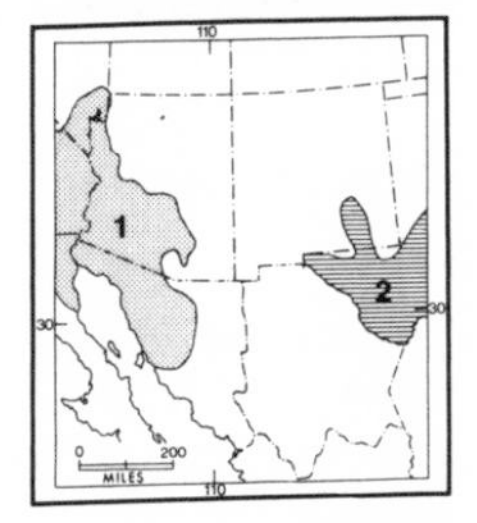

2. Mexican Ground Squirrel

Golden-mantled Ground Squirrel

Spermophilus lateralis

Order Rodentia **Family Sciuridae**

Identifying Features

A montane inhabiting ground squirrel with a white stripe running from the shoulder to the hip on each side of the back. The short tail is fully haired, edged with white and gray to yellowish below. Most commonly confused with chipmunks but differ in pattern of stripes. Unlike chipmunks these have no stripes on the head.

Measurements

Total length, 10.8 inches (275 mm); tail, 3.7 inches (95 mm); hind foot, 1.7 inches (42 mm); ear, 0.8 inch (20 mm); weight, 7 ounces (200 g).

Habitat

Generally in meadows or forest glades in higher mountains.

Life Habits

Feeds on plant material including buds, young leaves, flowers, seeds, berries, nuts and fungi. Insects are also readily taken. A single litter of four to eight, usually six, young is born in late spring. By early fall the adults and young have accumulated large amounts of body fat. Hibernation for five to six months is usual, especially at higher elevations.

Related Species

In the southern part of Chihuahua a related species occurs: Sierra Madre Ground Squirrel (*Spermophilus madrensis*).

1. Golden-mantled
Ground Squirrel

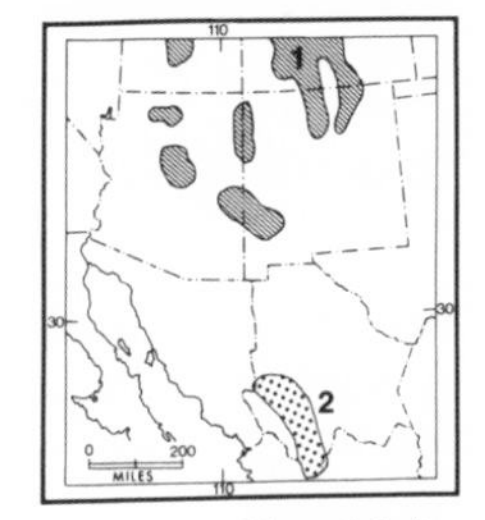

2. Sierra Madre
Ground Squirrel

Gunnison's Prairie Dog

Cynomys gunnisoni

Order Rodentia **Family Sciuridae**

Identifying Features

A large, ground squirrel-shaped rodent with a chunky body and a short tail. Pelage color a uniform light cinnamon buff.

Measurements

Total length, 13.8 inches (350 mm); tail, 2.4 inches (60 mm); hind foot, 2.2 inches (55 mm); ear, 0.5 inch (12 mm); weight, 2 pounds (900 g).

Habitat

Generally occurs in areas of open grassland, especially where the soil is compact and well drained.

Life Habits

Prairie dogs usually live in colonies, often covering many acres and including up to 200 or more individuals. In the colonies a complex social structure exists, with various divisions (wards) existing. Complex communications involving sight, sound and odor help provide cues for maintaining their social system. Food consists of various plant materials, especially short grasses. Roots and bulbs or other plants as well as worms and insects are also eaten. Formerly widely distributed in rangelands, prairie dogs are now much reduced in numbers and distribution.

Related Species

Two other species of prairie dogs occur in part of this region: Black-tailed Prairie Dog (*Cynomys ludovicianus*) and Utah Prairie Dog (*Cynomys parvidens*).

1. Gunnison's Prairie Dog

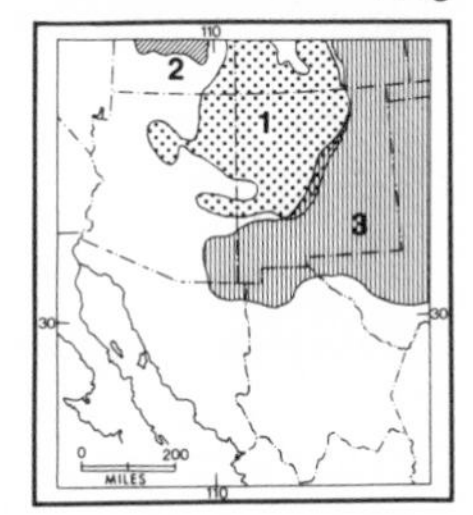

2. Utah Prairie Dog
3. Black-tailed Prairie Dog

Abert's Squirrel

Sciurus aberti

Order Rodentia **Family Sciuridae**

Identifying Features

A tree squirrel with characteristic large tufts of long hair on the ears and dark dorsal coat color.

Measurements

Total length, 21.6 inches (550 mm); tail, 10.2 inches (260 mm); hind foot, 2.8 inches (70 mm); ear, 1.5 inches (38 mm); weight, 1 pound 8 ounces (680 g).

Habitat

Restricted to ponderosa pine forests in mountainous areas.

Life Habits

Active throughout the year, this squirrel feeds on plant materials including leaf buds, flowers, herbs, fungi, acorns and especially the seeds of ponderosa pines. One, sometimes two, litters are born each year, one as early as May, the second as late as September. Three or four young per litter is common.

Related Species

Some authorities recognize the isolated population of squirrels on the North Rim of the Grand Canyon as a separate species: Kaibab Squirrel (*Sciurus kaibabensis*). The Kaibab Squirrel is here considered a subspecies of Abert's.

Abert's Squirrel

Nayarit Squirrel

Sciurus nayaritensis

Order Rodentia **Family Sciuridae**

Identifying Features

A large tree squirrel with underparts washed with yellowish color.

Measurements

Total length, 22 inches (565 mm); tail, 10.6 inches (270 mm); hind foot, 2.8 inches (70 mm); ear, 1.3 inches (34 mm); weight, 1 pound 12 ounces (800 g).

Habitat

The oak-pine zone of the Sierra Madre mountains and adjacent ranges.

Life Habits

Similar to the Abert's Squirrel.

Nayarit Squirrel

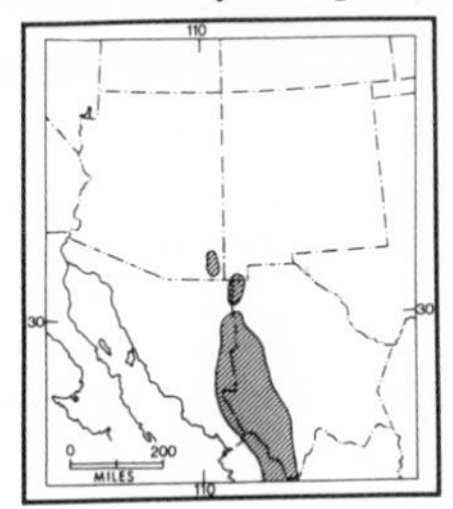

Arizona Gray Squirrel

Sciurus arizonensis

Order Rodentia **Family Sciuridae**

Identifying Features

Similar to Abert's Squirrel but with no ear tufts. Differs from the Nayarit Squirrel in that the belly is white. Differs from the Rock Squirrel in that the tail is clearly edged in white and the back is uniform in color.

Measurements

Total length, 21 inches (535 mm); tail, 9.8 inches (250 mm); hind foot, 2.8 inches (70 mm); ear, 1.2 inches (30 mm); weight, 1 pound 7 ounces (650 g).

Habitat

Generally in canyons and rim country where oaks, walnuts and some pine are present.

Life Habits

Similar to Abert's Squirrel. Food includes a wide variety of vegetable material including nuts, fruits, bark, berries, flowers and fungi. Much food, especially various nuts, is gathered and stored in hollow trees or in shallow holes in the ground cover of leaves and humus.

Arizona Gray Squirrel

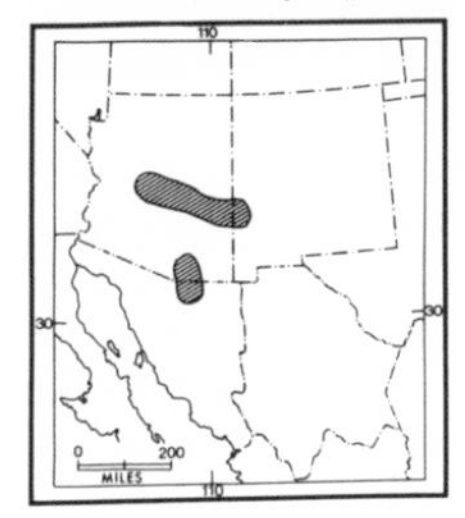

Red Squirrel

Tamiasciurus hudsonicus

Order Rodentia **Family Sciuridae**

Identifying Features

A small tree squirrel with a prominent black stripe along the side. The tail is narrow, shorter than the body and has a black edge and tip.

Measurements

Total length, 12.8 inches (325 mm); tail, 4.9 inches (125 mm); hind foot, 1.9 inches (49 mm); ear, 1 inch (25 mm); weight, 8.1 ounces (230 g).

Habitat

High montane situations, generally associated with spruce forests above 7,000 feet (2100 m) in elevation.

Life Habits

Similar to those of other tree squirrels.

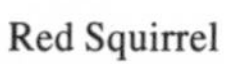
Red Squirrel

110
30
30
0
200
MILES
110
Sandy Truett

Botta's Pocket Gopher

Thomomys bottae

Order Rodentia **Family Geomyidae**

Identifying Features

Rarely seen above ground, this medium-sized rodent is well modified for life underground. Forefeet have long claws for digging; ears tiny; neck short; head wide, heavy and wedge-shaped; tail short, almost naked, with tactile hairs. Characteristic mounds of dirt are the best indicator of the presence of gophers.

Measurements

Total length, 9.5 inches (240 mm); tail, 3 inches (75 mm); hind foot, 1.3 inches (32 mm); ear, 0.4 inch (10 mm); weight, 6.7 ounces (190 g).

Habitat

Inhabit areas of soft soil. In the desert are most common along streams but are more widely distributed in grasslands.

Life Habits

Gophers feed entirely on vegetation. Some feeding is done above ground, but most occurs in a long feeding tunnel excavated under the ground. Much of their food is obtained by feeding on roots and bulbs, but it is not unusual for whole plants to be pulled down into a burrow beneath a plant. A nest is built in a deep tunnel, often under a rock or the roots of a bush or tree. Generally only one individual inhabits a tunnel system. One or two litters per year with two to ten, usually five, young per litter common.

Related Species

In general, one area has only one kind of pocket gopher. Differences in tooth and skull structure, size and pelage distinguish the five related species: Northern Pocket Gopher *(Thomomys talpoides);* Southern Pocket Gopher *(Thomomys umbrinus);* Plains Pocket Gopher *(Geomys bursarius);* Desert Pocket Gopher *(Geomys arenarius);* and Yellow-faced Pocket Gopher *(Pappogeomys castanops).*

1. Botta's Pocket Gopher

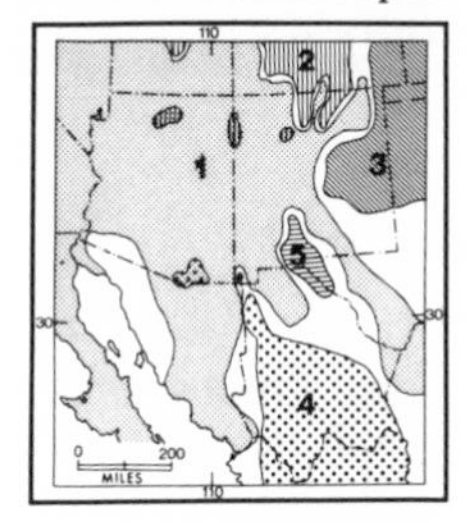

2. Northern Pocket Gopher
3. Plains Pocket Gopher
4. Southern Pocket Gopher
5. Desert Pocket Gopher

Yellow-faced Pocket Gopher

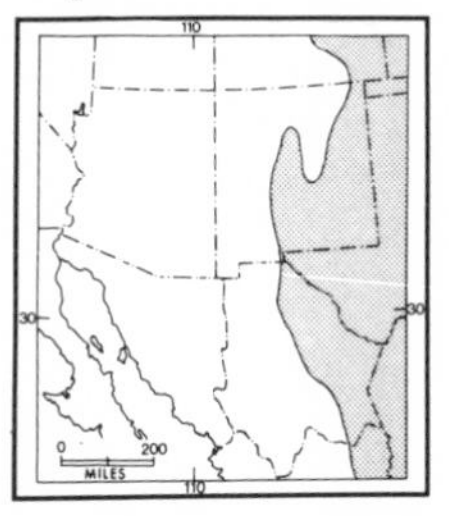

Silky Pocket Mouse

Perognathus flavus

Order Rodentia **Family Heteromyidae**

Identifying Features

Small mouse with soft fur, a wide head, short neck and tiny ears. The tail is untufted and shorter than the head and body. Color is generally light pink above and white below. Well developed fur-lined cheek pouches.

Measurements

Total length, 3.9 inches (100 mm); tail, 1.8 inches (45 mm); hind foot, 0.6 inch (16 mm); ear, 0.2 inch (6 mm); weight, 0.3 ounce (8 g).

Habitat

Common in short grass areas, often associated with sandy soil.

Life Habits

Food consists of seed of grasses and herbs, insects and green plant material. Active at night, almost always in late spring, summer and early fall. Cold seasons are spent underground, generally in hibernation. Three to six, usually four, young are born in a litter. Two or more litters may be born in a year.

Related Species

Plains Pocket Mouse *(Perognathus flavescens),* hind foot, 0.6 inch (16 mm); weight, 0.3 ounce (10 g); Little Pocket Mouse *(Perognathus longimembris),* hind foot, 0.7 in. (18 mm); weight 0.2 ounce (7 g); Arizona Pocket Mouse *(Perognathus amplus),* hind foot, 0.7 in. (18 mm); weight, 0.4 ounce (13 g); Great Basin Pocket Mouse *(Perognathus parvus),* hind foot, 0.9 in. (23 mm); weight 0.8 ounce (23 g); Long-tailed Pocket Mouse *(Perognathus formosus),* hind foot, 1 inch (24 mm); weight, 0.6 ounce (17 g).

1. Silky Pocket Mouse

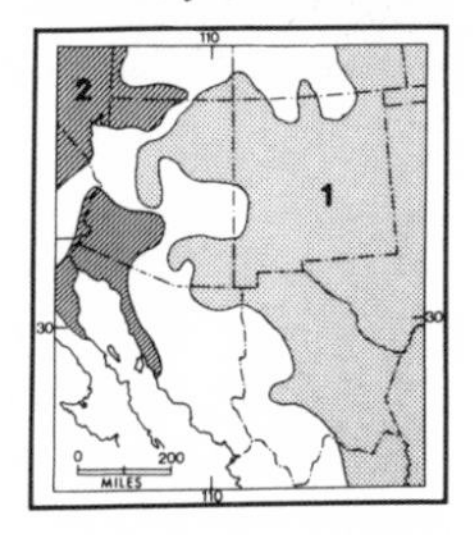

2. Little Pocket Mouse

1. Arizona Pocket Mouse
2. Plains Pocket Mouse
3. Great Basin Pocket Mouse

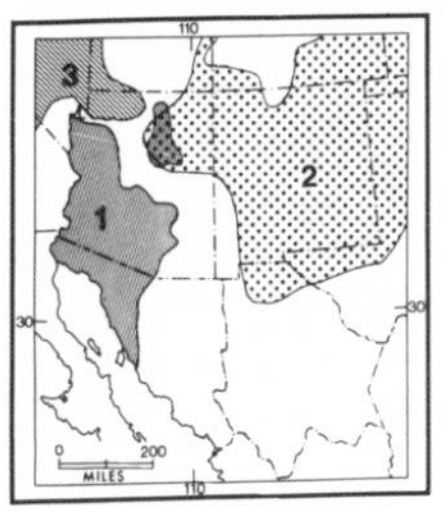

Long-tailed Pocket Mouse

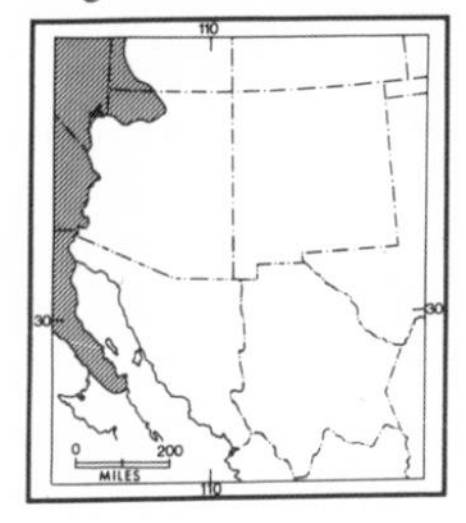

Spiny Pocket Mouse

Perognathus spinatus

Order Rodentia **Family Heteromyidae**

Identifying Features

Somewhat like the Silky Pocket Mouse but larger and with pelage darker colored and not so soft. Tail generally longer than the head and body and terminating in a weakly developed tuft.

Measurements

Total length, 8.2 inches (210 mm); tail, 4.5 inches (115 mm); hind foot, 1 inch (24 mm); ear, 0.4 inch (10 mm); weight, 0.9 ounce (25 g).

Habitat

Low desert floor in areas of sandy or gravelly soil.

Life Habits

Similar to those of the Silky Pocket Mouse. Food consists primarily of seeds, but a wide variety of green growth and insects is taken. Seeds are stored in the underground burrow system. Nocturnal activity is generally throughout the year but is reduced in cold periods. One or two litters per year with up to six young per litter.

Related Species

Six spiny pelaged pocket mice occur in this region, all in lower desert habitats. Differ in technical details. Bailey's Pocket Mouse (*Perognathus baileyi*), hind foot, 1 inch (25 mm); weight, 1 ounce (28 grams); Hispid Pocket Mouse (*Perognathus hispidus*), hind foot, 1 inch (26 mm); weight, 1.2 ounces (35 grams); Desert Pocket Mouse (*Perognathus penicillatus*), hind foot, 0.6 inch (23 mm); weight, 0.6 ounce (18 grams); Rock Pocket Mouse (*Perognathus intermedius*), hind foot, 0.8 inch (20 mm); weight, 0.6 ounce (16 grams); Nelson's Pocket Mouse (*Perognathus nelsoni*), hind foot, 1 inch (24 mm); weight, 0.7 ounce (20 grams).

1. Spiny Pocket Mouse

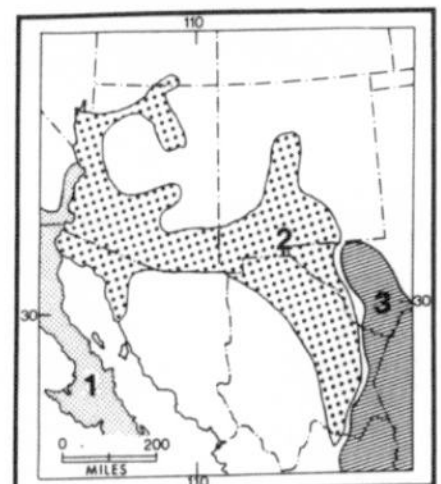

2. Rock Pocket Mouse
3. Nelson's Pocket Mouse

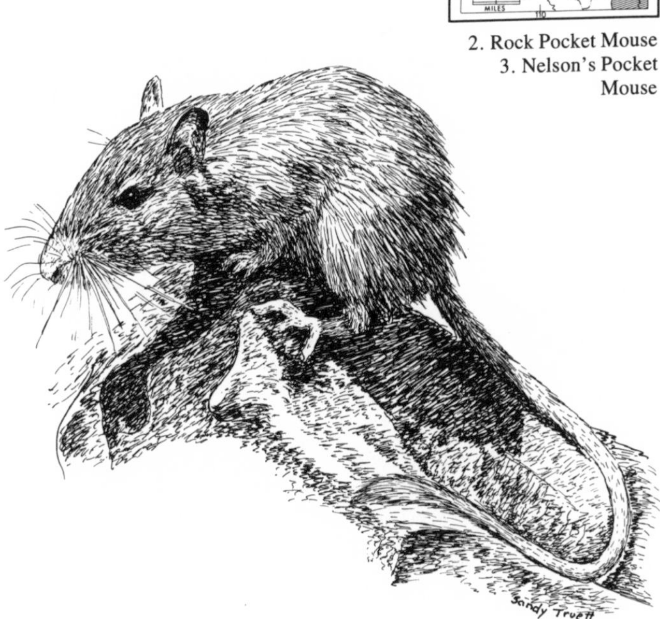

Desert Pocket Mouse

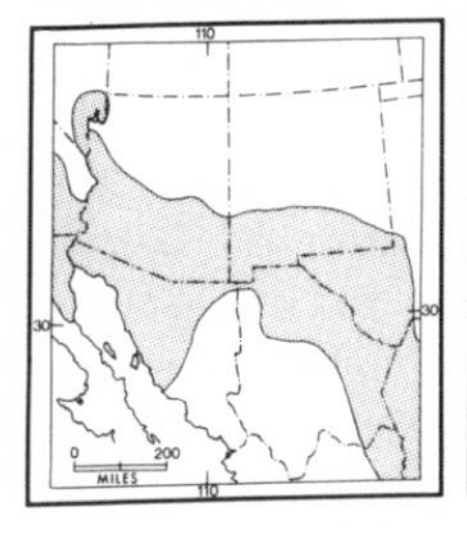

Bailey's Pocket Mouse

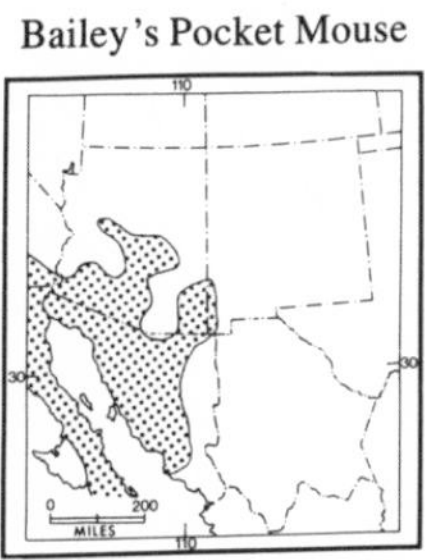

Hispid Pocket Mouse

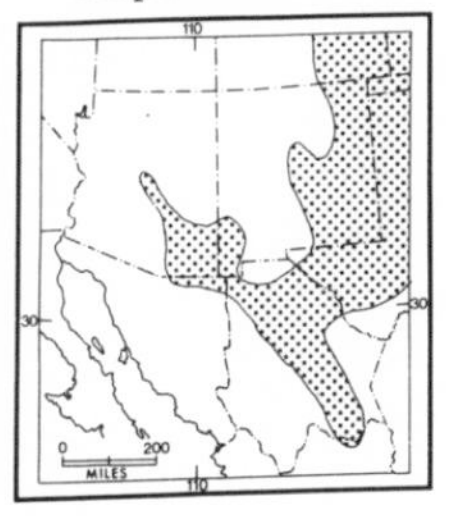

Merriam's Kangaroo Rat

Dipodomys merriami

Order Rodentia **Family Heteromyidae**

Identifying Features
The kangaroo rat has striking modifications for jumping (saltatorial) locomotion. Hind legs and feet are greatly enlarged; tail long, terminates in a tuft. Prominent white rump stripe across thighs. Dorsal color generally tan, buff or cinnamon; belly white.

Measurements
Total length, 9.8 inches (250 mm); tail, 5.9 inches (150 mm); hind foot, 1.5 inches (39 mm); ear, 0.5 inch (14 mm); weight, 1.6 ounces (44 g).

Habitat
Inhabits grasslands to low desert, all in relatively open microhabitats where jumping is useful.

Life Habits
Nocturnal, seed-feeding rodents that store excess food in shallow holes or underground burrows. When available, insects and newly sprouted seed are eaten. Active throughout the year. Generally two or three young are born in a litter. Two litters a year are common, in winter and in the summer.

Related Species
Ord's Kangaroo Rat *(Dipodomys ordii),* hind foot, 1.6 inches (40 mm); weight, 2.5 ounces (70 grams); Chisel-toothed Kangaroo Rat *(Dipodomys microps),* hind foot, 1.6 inches (41 mm); weight, 2.3 ounces (64 grams); Banner-tailed Kangaroo Rat *(Dipodomys spectabilis),* hind foot, 2 inches (50 mm); weight, 4.2 ounces (120 grams); Nelson's Kangaroo Rat *(Dipodomys nelsoni),* hind foot, 1.9 inches (48 mm); weight, 3.9 ounces (110 grams); Desert Kangaroo Rat *(Dipodomys deserti),* hind foot, 2.2 inches (55 mm); weight, 4.2 ounces (120 grams).

Merriam's Kangaroo Rat

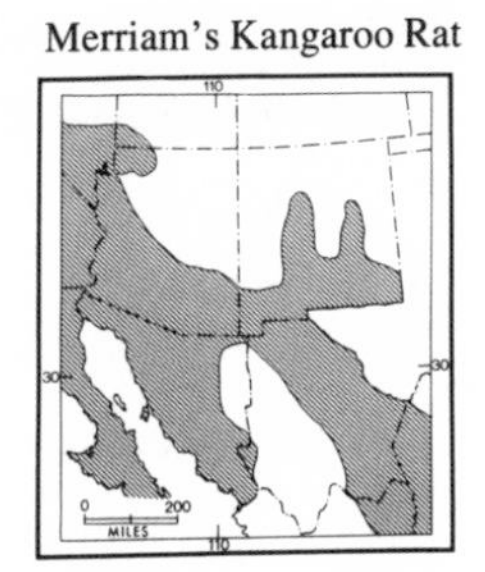

1. Banner-tailed Kangaroo Rat
2. Nelson's Kangaroo Rat
3. Chisel-toothed Kangaroo Rat

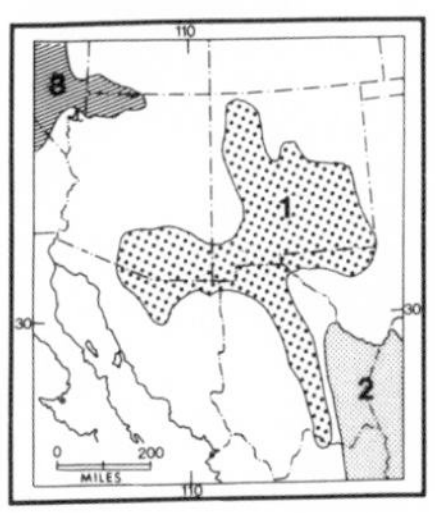

Ord's Kangaroo Rat

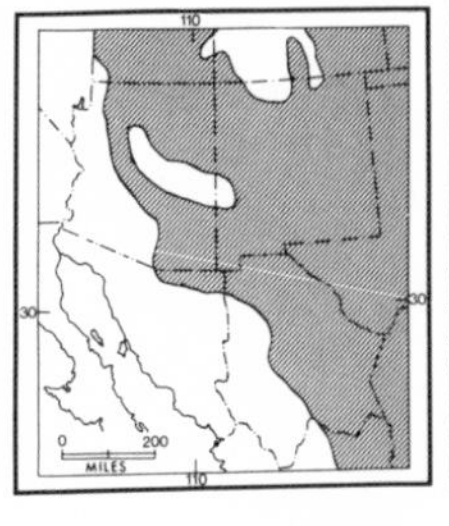

Desert Kangaroo Rat

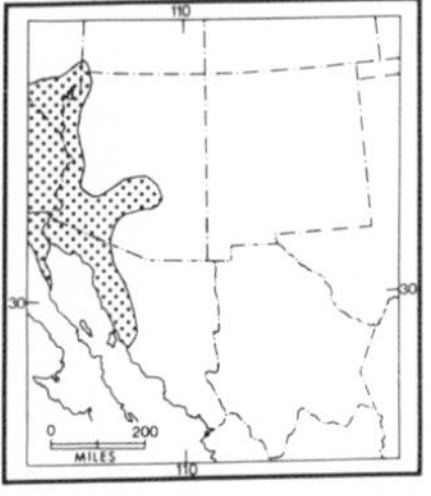

Beaver

Castor canadensis

Order Rodentia **Family Castoridae**

Identifying Features
The beaver is a unique rodent characterized by large size and semi-aquatic adaptations including a large, flattened, scale-covered tail. The hind feet are webbed and the ears are small. Pelage with dense underfur and many long coarse guard hairs.

Measurements
Total length, 36 inches (980 mm); tail, 16 inches (400 mm); hind foot, 6.7 inches (170 mm); ear, 1.3 inches (33 mm); weight, 60 pounds (27 kg).

Habitat
In or along permanent streams bordered by trees.

Life Habits
Generally living in family groups in or along permanent streams, beavers feed on the bark and cambium layer of various bushes and trees, especially aspens, birches, and willows. They often build a dam of sticks, rocks and mud, making a pond of a small stream. In such situations they usually construct a dome-shaped lodge of sticks and mud in the pond that has an underwater entrance. In some places ''lodges'' are constructed in stream banks. A litter of two to eight, usually four, kits is born in April or May.

Beaver

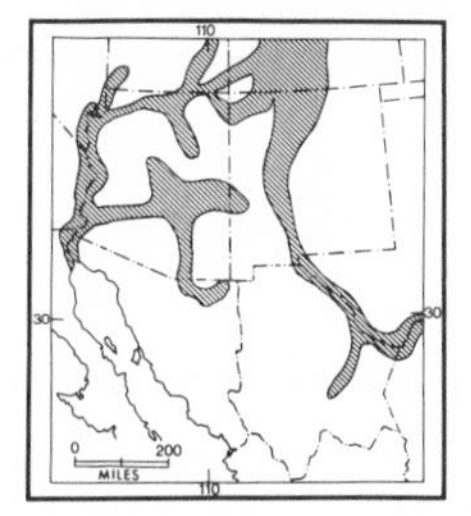

Western Harvest Mouse

Reithrodontomys megalotis

Order Rodentia **Family Cricetidae**

Identifying Features

A relatively small mouse with smallish ears, and a long slender, sparsely-haired tail. Not very evident but unique among other members of this family that live in this region is the presence of a vertical groove on the anterior face of each upper incisor.

Measurements

Total length, 5.5 inches (140 mm); tail, 2.6 inches (65 mm); hind foot, 0.7 inch (17 mm); ear, 0.5 inch (12 mm); weight, 0.4 ounce (12 g).

Habitat

Grassy areas, being most numerous in desert grasslands and mountain meadows.

Life Habits

A generally nocturnal rodent that feeds on a variety of seeds and plant growth. A nest of grass at ground level or sometimes in a bush is generally used as a retreat during the day. The nest is sometimes a modification of an abandoned bird nest. After a gestation period of 23 days a litter of one to seven, usually four, young is born. The breeding season is in the spring and summer from April through October.

Related Species

Three related species occur in the area: Plains Harvest Mouse (*Reithrodontomys montanus*), Burt's Harvest Mouse (*Reithrodontomys burti*), and Fulvous Harvest Mouse (*Reithrodontomys fulvescens*).

Western Harvest Mouse

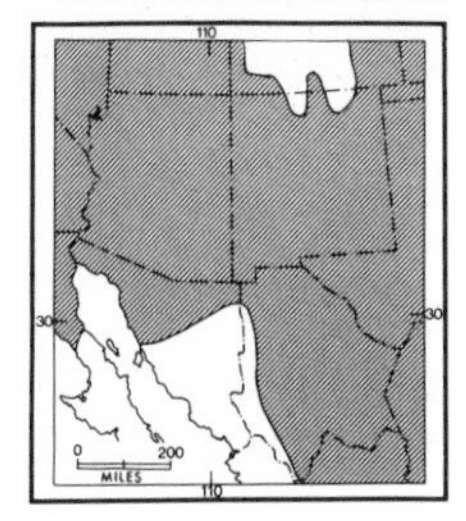

1. Plains Harvest Mouse
2. Burt's Harvest Mouse

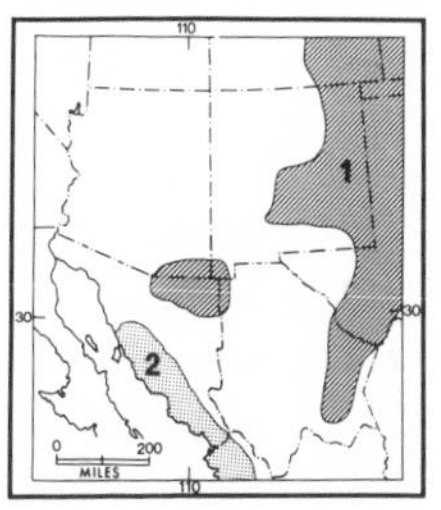

Fulvous Harvest Mouse

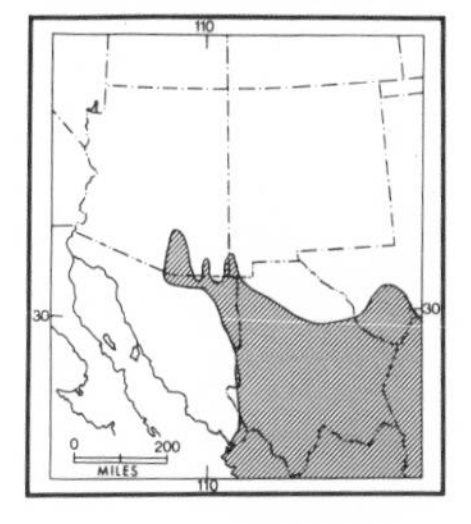

White-footed Mouse

Peromyscus leucopus

Order Rodentia **Family Cricetidae**

Identifying Features

Similar to the Western Harvest Mouse but larger, especially the ears and the diameter of the tail.

Measurements

Total length, 7.1 inches (180 mm); tail, 3.1 inches (80 mm); hind foot, 0.9 inch (22 mm); ear, 0.7 inch (17 mm); weight, 1.1 ounces (30 g).

Habitat

Generally occurs in woodland areas below the pine forests but some related species occurs in almost all habitats.

Life Habits

Similar to the Western Harvest Mouse but somewhat more omnivorous. Seeds, buds, berries and fruits make up much of the food but worms, insects and other small animals are also eaten. After a gestation of about 28 days a litter of three to seven, usually four, young is born. Sexual maturity is reached in about six weeks. Most young are born between March and September.

Related Species

Related species occupy such microhabitats as rock bluffs, forests, grasslands, and deserts, from sea level to above tree line. They rarely occur in cultivated fields or cities. Differences are relatively minor and generally require comparative material. The nine related species that occur in this region are: Canyon Mouse *(Peromyscus crinitus);* Merriam's Mouse *(Peromyscus merriami);* Cactus Mouse *(Peromyscus eremicus);* Deer Mouse *(Peromyscus maniculatus);* Brush Mouse *(Peromyscus boylii);* White-ankled Mouse *(Peromyscus pectoralis);* Piñon Mouse *(Peromyscus truei);* Rock Mouse *(Peromyscus difficilis);* and Black-eared Mouse *(Peromyscus melanotis)*.

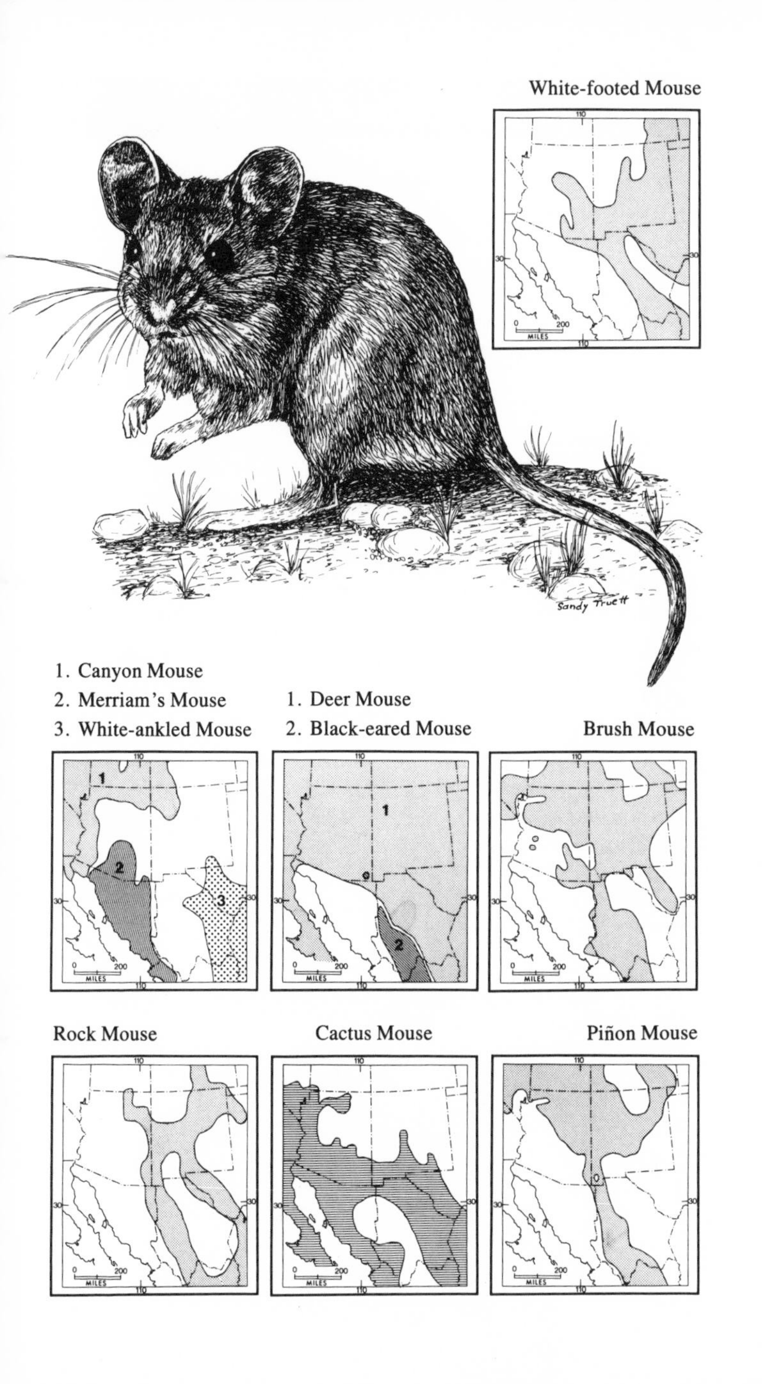
White-footed Mouse
Sandy Truett
1. Canyon Mouse
2. Merriam's Mouse
3. White-ankled Mouse
1. Deer Mouse
2. Black-eared Mouse
Brush Mouse
Rock Mouse
Cactus Mouse
Piñon Mouse
110
30
0
200
MILES
1
2
3

Northern Pygmy Mouse

Baiomys taylori

Order Rodentia **Family Cricetidae**

Identifying Features

Similar to the Western Harvest Mouse but smaller and with a very short, hair-covered tail. The pelage color is dark gray, soft and relatively shaggy.

Measurements

Total length, 4.1 inches (105 mm); tail, 1.8 inches (45 mm); hind foot, 0.6 inch (14 mm); ear, 0.4 inch (11 mm); weight, 0.4 ounce (10 g).

Habitat

Restricted to taller grasslands along the Chihuahuan Desert edge.

Life Habits

The Pygmy Mouse feeds primarily on grass shoots and other green vegetative material. They construct small runways in the grass where the tips of cut grass leaves can often be seen. Litter size is relatively small, ranging from one to three with two most common.

Northern Pygmy Mouse

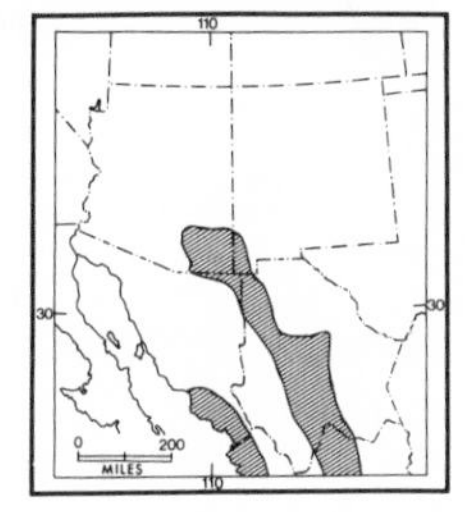

Southern Grasshopper Mouse

Onychomys torridus

Order Rodentia **Family Cricetidae**

Identifying Features

A plump-bodied mouse with a short, thick tail that is constricted at the base and is only about half the length of the head and body. The back is generally pale cinnamon to light brown in color. The belly and feet are white.

Measurements

Total length, 6.3 inches (106 mm); tail, 2.4 inches (60 mm); hind foot, 0.9 inch (22 mm); ear, 0.7 inch (19 mm); weight, 1.2 ounces (35 g).

Habitat

Generally found in sandy vegetated areas in the desert and desert grasslands.

Life Habits

Nocturnal, active throughout the year. Food consists of a variety of plant and animal material. This species is much more carnivorous than other rodents in the region. In a given area fewer individuals are present than are characteristic of pocket mice or deer mice, but the grasshopper mice have much larger feeding areas. Four or five young per litter appears to be common.

Related Species

Two related species occur in the area. Both are very similar to this species: Northern Grasshopper Mouse *(Onychomys leucogaster)* and Mearns' Grasshopper Mouse *(Onychomys arenicola)*, known only from southwestern New Mexico.

Southern Grasshopper
Mouse

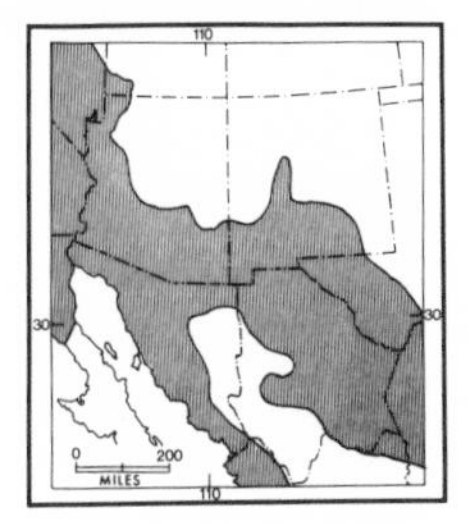

Northern Grasshopper
Mouse

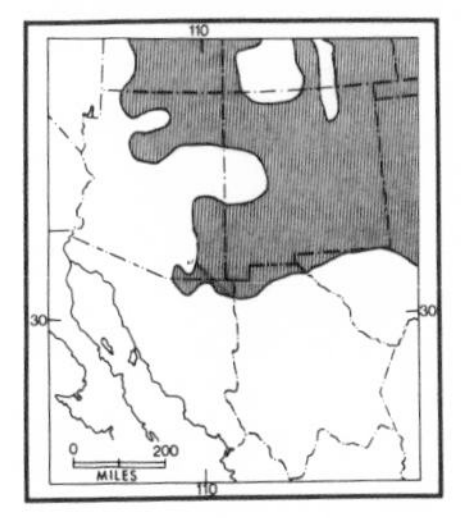

Hispid Cotton Rat

Sigmodon hispidus

Order Rodentia **Family Cricetidae**

Identifying Features

Small rat-sized rodent with tail shorter than head and body. Pelage coarsely grizzled, generally some mixture of blackish or brownish and buffy or grayish hairs. The sides are slightly lighter than the back.

Measurements

Total length, 9.8 inches (250 mm); tail, 4.1 inches (105 mm); hind foot, 1.3 inches (32 mm); ear, 0.7 inch (17 mm); weight, 3.2 ounces (90 g).

Habitat

Restricted to areas of tall grasses and weed-grown areas.

Life Habits

Food consists primarily of green vegetation including stems and blades of grasses and stalks of weeds. They are active throughout the year. Two to ten young are born in a litter with five or six most common. Young are born during the warm months between June and October or November.

Related Species

Cotton rats are essentially inhabitants of tropical and subtropical grasslands. Some occur as far north as the central United States. Three related species occur in this area: Arizona Cotton Rat (*Sigmodon arizonae*); Tawny-bellied Cotton Rat (*Sigmodon fulviventer*); and Yellow-nosed Cotton Rat (*Sigmodon ochrognathus*).

1. Hispid Cotton Rat

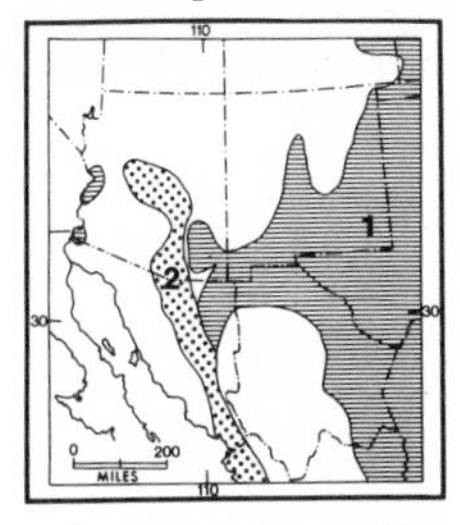

2. Arizona Cotton Rat

Yellow-nosed
Cotton Rat

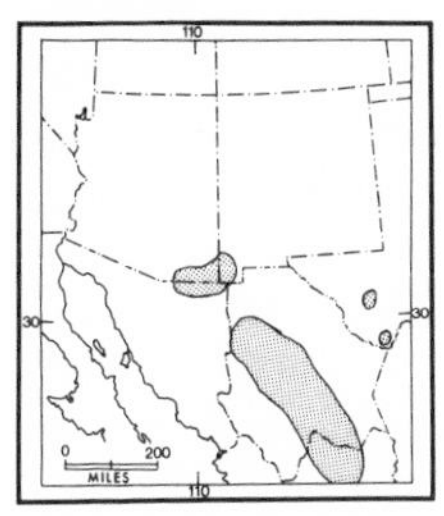

Tawny-bellied
Cotton Rat

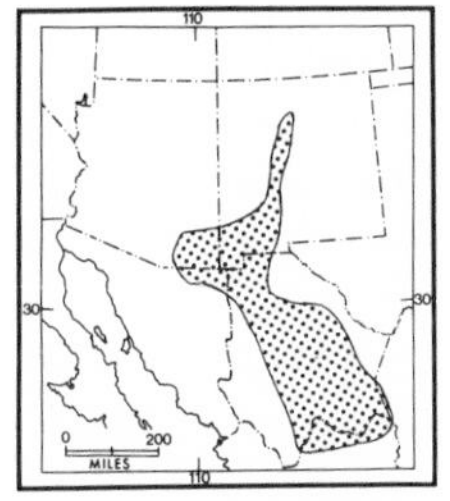

White-throated Woodrat

Neotoma albigula

Order Rodentia **Family Cricetidae**

Identifying Features

Large rats with tail shorter than the head and body, thick, round and blunt-ended, covered with short hairs and bi-colored, dark above, white below. The ears are large and naked. The pelage is soft, dense, dark above and white ventrally. Wood rats often build characteristic piles of sticks and other material around nest sites.

Measurements

Total length, 15 inches (390 mm); tail, 5.9 inches (150 mm); hind foot, 1.5 inches (37 mm); ear, 1.4 inches (35 mm); weight, 6.3 ounces (180 g).

Habitat

Live in a wide range of natural habitats from forest edge, rock ledges and brush areas to low desert habitats. Usually occur near *Opuntia* cacti (prickly pear and cholla).

Life Habits

Active throughout the year. Food consists of a wide variety of herbaceous materials including new stem and leaf shoots of trees and, especially, parts of various cacti, especially *Opuntia*. Breeding occurs from April to September with one to three young in the litter. Dens range from a few sticks near the grass nest concealed in a rock crevice to extensive mounds of sticks, cactus sections and other debris in open areas.

Related Species

Woodrats are most common in the southwestern United States. In this region six related species are found differing in minor details of structure and size: Southern Plains Woodrat *(Neotoma micropus);* Desert Woodrat *(Neotoma lepida);* Stephens' Woodrat *(Neotoma stephensi);* Mexican Woodrat *(Netotoma mexicana);* Goldman's Woodrat *(Neotoma goldmani);* and Bushy-tailed Woodrat *(Neotoma cinerea)*.

White-throated
Woodrat

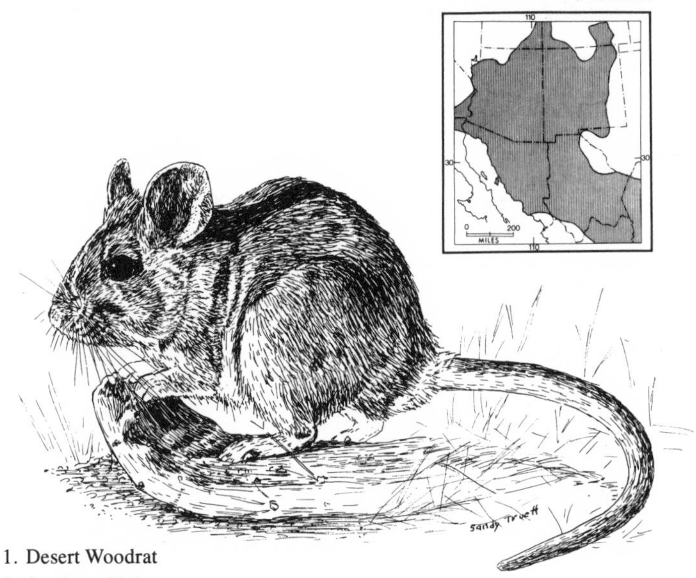

1. Desert Woodrat
2. Southern Plains Woodrat

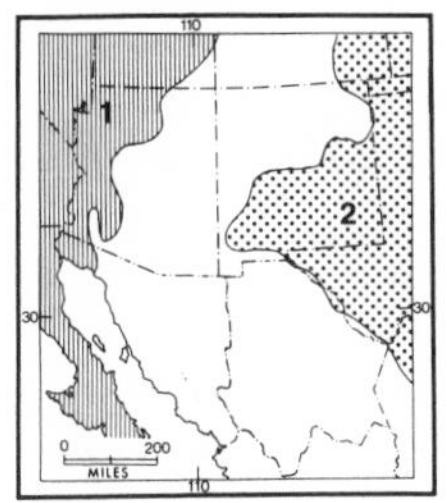

Stephens' Woodrat

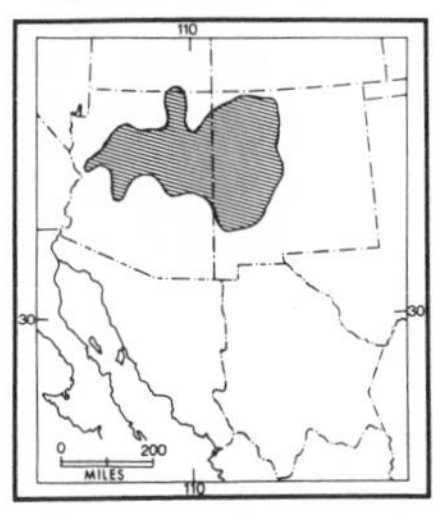

1. Mexican Woodrat
2. Goldman's Woodrat

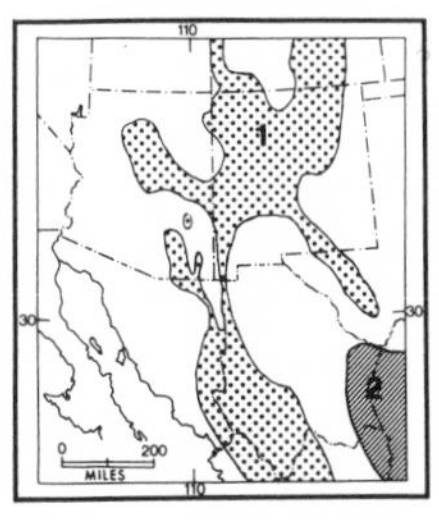

Bushy-tailed Woodrat

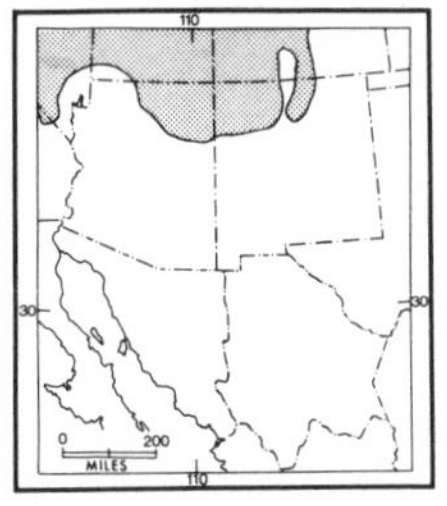

Southern Red-backed Vole

Clethrionomys gapperi

Order Rodentia **Family Cricetidae**

Identifying Features

Chunky, extremely short-tailed mice with short ears and long, loose grizzled pelage. A dorsal strip ranging from bright chestnut to yellowish brown, even black, gives rise to the common name.

Measurements

Total length, 5.3 inches (135 mm); tail, 1.6 inches (40 mm); hind foot, 0.7 inch (19 mm); ear, 0.5 inch (13 mm); weight, 1 ounce (25 g).

Habitat

Montane forests generally above 8,000 feet (2400 m) in elevation.

Life Habits

Active throughout the year. During the winter they construct feeding tunnels under the snow. Activity is not restricted to the night. Food is almost entirely vegetation, with a leaf mold growing fungus, *Endogone*, making up all of the food of some individuals. Other foods include green vegetation and sprouting seeds. Three to eight, usually four or five, young are born in a litter and more than one litter each summer is common.

Related Species

Voles are primarily Arctic grass-eating rodents that occur in the higher mountain habitats of the Southwest. Five related species differing in minor details occur in the region: Heather Vole (*Phenacomys intermedius*); Meadow Vole (*Microtus pennsylvanicus*); Montane Vole (*Microtus montanus*); Long-tailed Vole (*Microtus longicaudus*); and Mexican Vole (*Microtus mexicanus*).

Southern Red-backed Vole

Meadow Vole

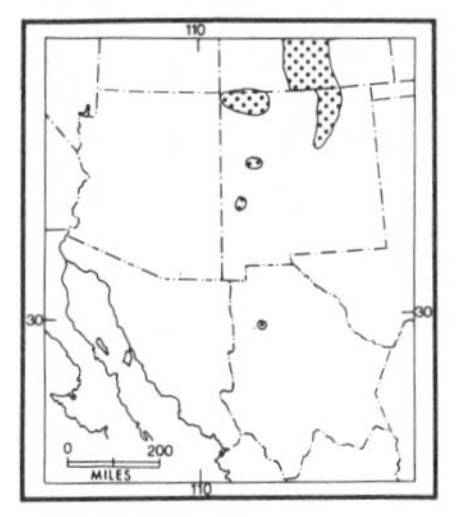

Montane Vole

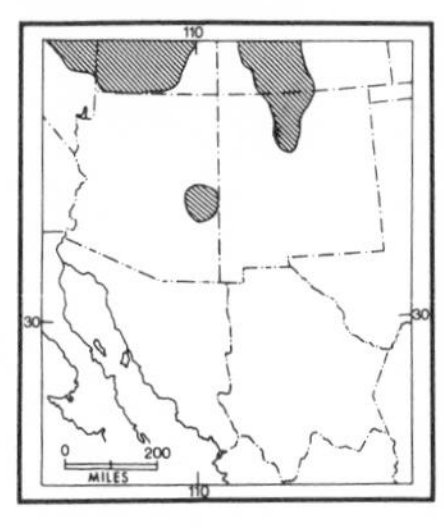

Long-tailed Vole

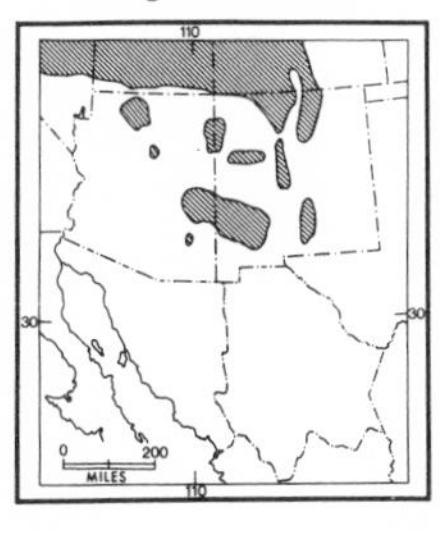

1. Heather Vole
2. Mexican Vole

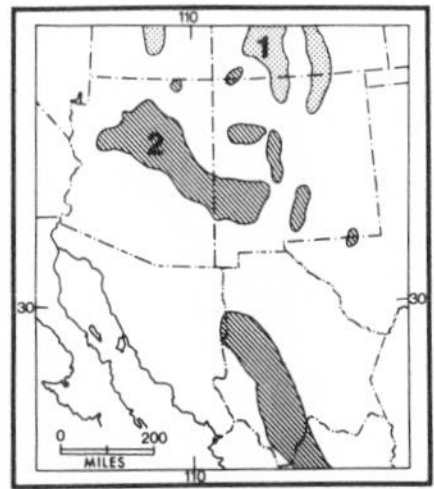

Muskrat

Ondatra zibethicus

Order Rodentia **Family Cricetidae**

Identifying Features
A large rodent with a scaly, laterally compressed tail that is about the length of the head and body.

Measurements
Total length, 20 inches (500 mm); tail, 10 inches (250 mm); hind foot, 2.8 inches (72 mm); ear, 0.8 inch (21 mm); weight, 4 pounds (1.8 kg).

Habitat
Permanent streams, lakes and ponds up to elevations of 11,000 feet (3300 m), generally in still water habitats.

Life Habits
Muskrats are active throughout the year, usually at night. Food consists of aquatic and riparian plants. When adjacent to fields, muskrats occasionally eat cultivated crops such as corn. Dens are constructed as domed structures rising above water with the entrance below the water line. In other situations, dens are constructed in the banks of streams or lakes, again utilizing underwater entrances. One to eleven young are born in a litter with six being the usual number. At high elevations and in the north two litters per year are common, but at low elevations in the south three and even four litters may be produced.

Muskrat

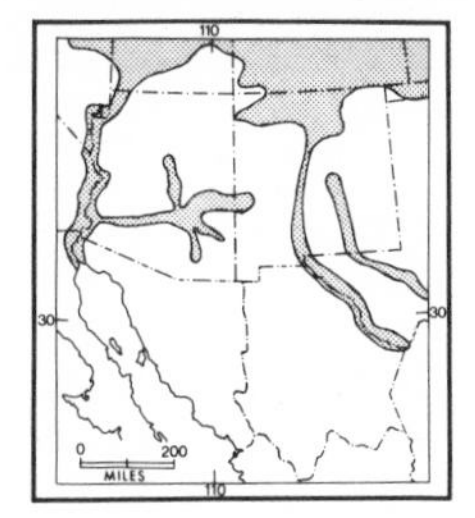

Black Rat

Rattus rattus

Order Rodentia **Family Muridae**

Identifying Features

Resembles the woodrat in general size but differs in having short ears, a thinly haired scaly tail that is longer than the head and body, a dark grayish back and grayish belly and feet.

Measurements

Total length, 14.6 inches (370 mm); tail, 7.7 inches (195 mm); hind foot, 1.4 inches (36 mm); ear, 0.9 inch (22 mm); weight, 7 ounces (200 g).

Habitat

Generally associated with man. Often found in buildings, trash heaps, city dumps and similar situations. Rarely found in native habitats.

Life Habits

This rat is a native of the Old World and a commensal of man. Rarely does it occur far from man's activities and structures. Buildings, canals, and cultivated fields are all typical habitats. When food is available, the reproductive rate of these rats is high. Sexual maturity can occur in three to four months and, after a gestation period of only 22 days, a litter of 6 to 22 young can be produced.

Related Species

The Norway Rat (*Rattus norvegicus*), also introduced from the Old World, has a shorter tail and small ears. The white and black rats sold in pet stores and as laboratory animals are members of this species.

Sandy Truett

House Mouse

Mus musculus

Order Rodentia **Family Muridae**

Identifying Features

Similar to the Western Harvest Mouse but with a long, scaly, hairless tail that is not white ventrally. Back, a dirty gray color. Belly is often grayish but sometimes has a whitish or buffy wash. From a side view the upper incisors have a characteristic notch in the grinding surface.

Measurements

Total length, 6.7 inches (170 mm); tail, 3.1 inches (80 mm); hind foot, 0.7 inch (18 mm); ear, 0.5 inch (13 mm); weight, 0.7 ounce (19 g).

Habitat

Generally in association with man and his agricultural areas. Rarely found in natural habitats.

Life Habits

Like the Black Rat, the House Mouse is a native of the Old World and a commensal of man. When food is available, this species is also capable of producing several large litters of young in a year.

House Mouse

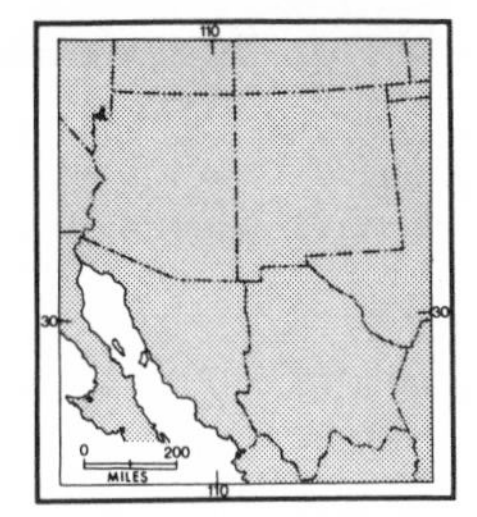

Western Jumping Mouse

Zapus princeps

Order Rodentia **Family Zapodidae**

Identifying Features

Similar to the White-footed Mouse but adapted for jumping locomotion. Hind feet and legs somewhat enlarged, tail very long. Like the Western Harvest Mouse, the anterior surface of the upper incisors is grooved.

Measurements

Total length, 9.5 inches (240 mm); tail, 5.7 inches (145 mm); hind foot, 1.3 inches (32 mm); ear, 0.6 inch (16 mm); weight, 0.8 ounce (23 g).

Habitat

Montane habitats between 6,000 and 11,000 feet (1800 to 3300 m), especially in aspen and in willows along streams.

Life Habits

This mouse hibernates during the winter after becoming extremely fat in the early fall. Food consists of seeds of various grasses. Only one litter per year is produced. These, ranging from two to seven in number, are born in early summer. Since young must reach adult size before they store fat for hibernation, they are usually active above ground after the adults have already begun hibernation.

Western Jumping Mouse

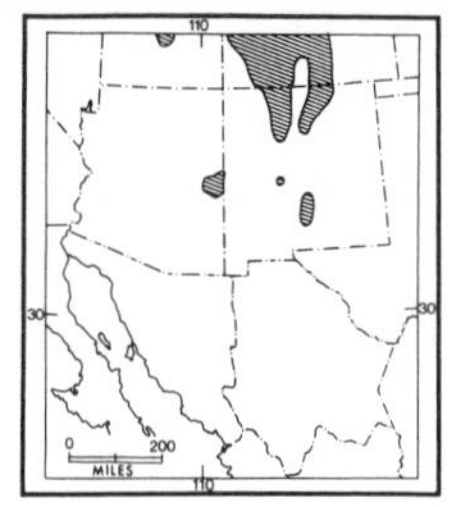

Porcupine

Erethizon dorsatum

Order Rodentia **Family Erethizontidae**

Identifying Features

A large rodent with unique specialized quills on the back and tail.

Measurements

Total length, 35 inches (875 mm); tail, 8.3 inches (210 mm); hind foot, 4.3 inches (110 mm); ear, 1.2 inches (30 mm); weight, 18 pounds (8.2 kg).

Habitat

Most common in coniferous forests but found in riparian situations, in desert grasslands and desert edge.

Life Habits

Feeds primarily on young twigs and leaves of a series of low plants. In winter they feed on pine needles, mistletoe and, to some extent, on the inner bark of trees, especially conifers. Activity is generally at night. One young, sometimes two, is born generally in some rock crevice or shallow cave. The gestation period is about four months, and the young are usually born in late spring.

Porcupine

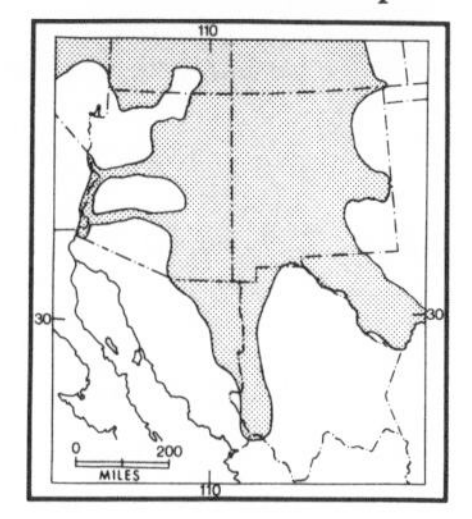

Eastern Cottontail

Sylvilagus floridanus

Order Lagomorpha **Family Leporidae**

Identifying Features

This is a medium-sized cottontail with relatively short ears and a characteristic rusty colored patch at the nape of the neck. Inner surface of ear has scattered hair.

Measurements

Total length, 16.5 inches (418 mm); tail, 2.2 inches (55 mm); hind foot, 3.5 inches (90 mm); ear, 2.4 inches (60 mm); weight, 3.3 pounds (1.5 kg).

Habitat

In most of this area, these cottontails are restricted to oak woodlands, especially in riparian situations.

Life Habits

Cottontails feed on various grasses, herbs and other green vegetation. They are most active in the early morning and late evening but shift the time of activity seasonally, becoming almost completely nocturnal during the hotter months. Young are born during the warmer months. Three or four litters with four to seven young in each are born each year. At birth the young are without hair, blind, and helpless. They are generally born in a fur-lined nest in a hole in the ground after a gestation period of about four weeks. Sexual maturity is reached nine to ten months after birth.

Eastern Cottontail

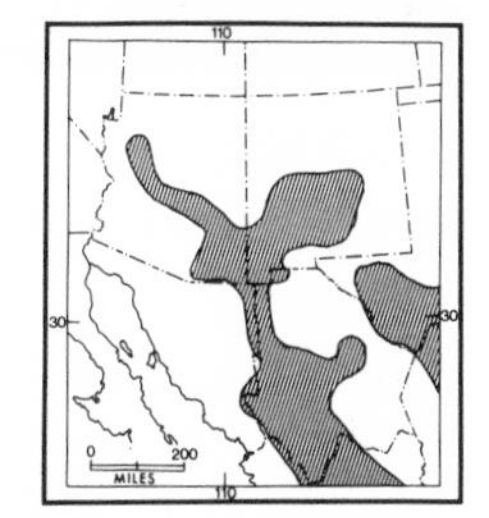

Nuttall's Cottontail

Sylvilagus nuttallii

Order Lagomorpha **Family Leporidae**

Identifying Features

Similar to the Eastern Cottontail but with inside of the ear densely furred. Occurs at high elevations and north of the range of the Eastern Cottontail.

Measurements

Total length, 15.6 inches (398 mm); tail, 1.3 inches (34 mm); hind foot, 4 inches (100 mm); ear, 2.4 inches (62 mm); weight, 2.5 pounds (1.1 kg).

Habitat

Generally above 6,000 feet (1800 m) in treeless, brush covered areas.

Life Habits

Generally similar to those of the Eastern Cottontail.

Nuttall's Cottontail

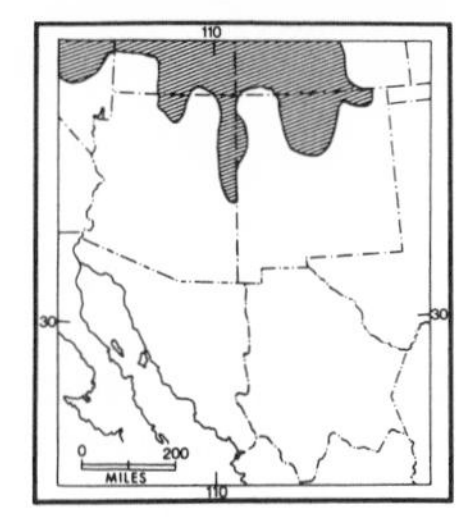

Desert Cottontail

Sylvilagus audubonii

Order Lagomorpha **Family Leporidae**

Identifying Features

Similar to the Eastern Cottontail but ears somewhat longer, color lighter and size generally slightly smaller. These are the cottontails of the low deserts.

Measurements

Total length, 15.4 inches (390 mm); tail, 1.7 inches (42 mm); hind foot, 3.8 inches (96 mm); ear, 2.7 inches (69 mm); weight, 1.8 pounds (850 g).

Habitat

Elevations below 6,000 feet (1800 m), generally in brushy areas.

Life Habits

Generally similar to those of the Eastern Cottontail.

Desert Cottontail

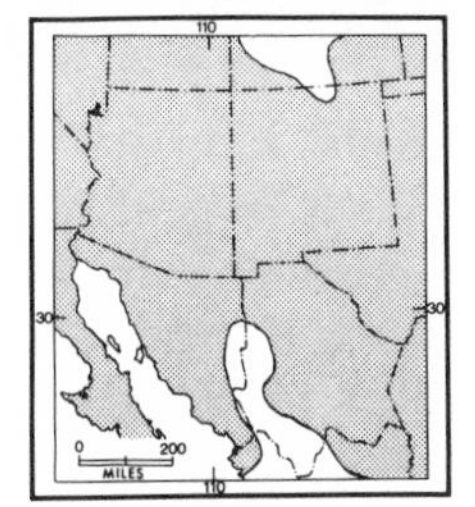

Black-tailed Jack Rabbit

Lepus californicus

Order Lagomorpha **Family Leporidae**

Identifying Features

The Black-tailed Jack Rabbit has large ears and greatly elongated hind legs. The dorsal surface of the tail is black.

Measurements

Total length, 22 inches (570 mm); tail, 3.1 inches (80 mm); hind foot, 5.3 inches (135 mm); ear, 4.5 inches (115 m); weight, 4.5 pounds (2.1 kg).

Habitat

Mainly at elevations below 6,000 feet (1800 m), often in brushy areas.

Life Habits

Food consists of various grasses and herbs. Feeding occurs most commonly at night. In the hot season they are most active at dusk and daybreak. The young are born fully haired and with their eyes open. They are able to follow the mother within a few minutes after birth. One to seven young (usually four) are born in a litter, and more than one litter per year appears normal.

Related Species

Snowshoe Hare (*Lepus americanus*) at elevations above 8,000 feet (2400 m); ears short, less than 3 inches (75 mm); White-tailed Jack Rabbit (*Lepus townsendii*) at elevations as low as 5,000 feet (1500 m) up to timberline; ears long as in Black-tailed Jack Rabbit; dorsal surface of tail white.

Black-tailed Jack
Rabbit

White-tailed Jack
Rabbit

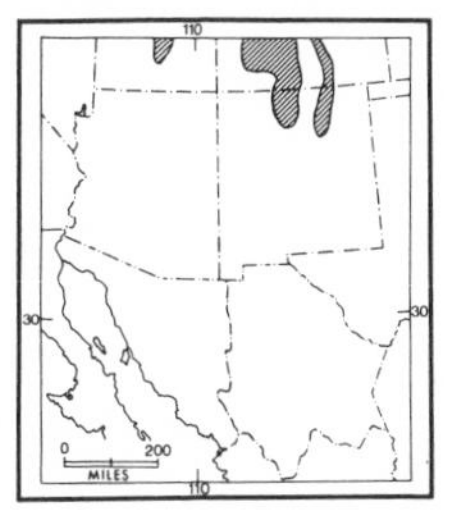

Antelope Jack Rabbit

Lepus alleni

Order Lagomorpha **Family Leporidae**

Identifying Features

Similar to the Black-tailed Jack Rabbit but much larger. This is the largest rabbit in the New World. It attempts to confuse predators by "flashing" its white undercoat while running.

Measurements

Total length, 30 inches (625 mm); tail, 2.6 inches (65 mm); hind foot, 5.3 inches (135 mm); ear, 7.1 inches (180 mm); weight, 9 pounds (4.1 kg).

Habitat

Restricted to the upper parts of the Sonoran Desert in south central Arizona southward into Sonora.

Life Habits

Similar to the Black-tailed Jack Rabbit. The number of young per litter is one to five with two being the most common number. Three to four litters per year are produced.

Related Species

White-sided Jack Rabbit (*Lepus callotis*), smaller and in the Chihuahuan Desert.

1. Antelope Jack Rabbit

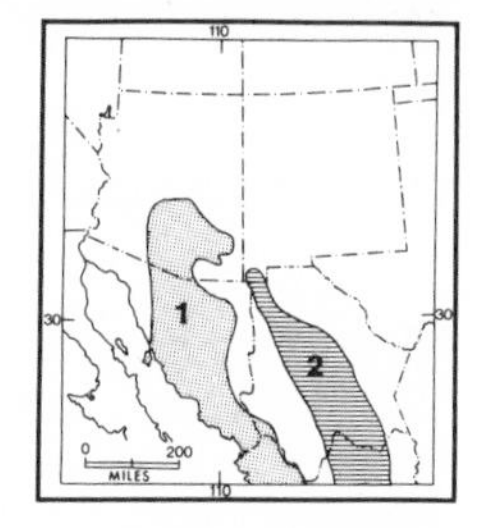

2. White-sided Jack Rabbit

Ghost-faced Bat

Mormoops megalophylla

Order Chiroptera **Family Mormoopidae**

Identifying Features

This bat lacks a leaflike growth on the nose but does have a complex area of folded skin on the lower lip. The ears are short and broad. The interfemoral membrane (between the legs) is long and the tail is only about half the length of the interfemoral membrane. Its tip is free on the dorsal surface of the membrane. Color is either dark reddish brown or light cinnamon brown.

Measurements

Total length, 3.7 inches (95 mm); tail, 1 inch (25 mm); hind foot, 0.5 inch (12 mm); ear, 0.6 inch (15 mm); forearm length, 2 inches (53 mm); weight 0.5 ounce (14 g).

Habitat

Essentially a tropical species, this bat occurs northward in areas of low desert shrub. Day roosts are in moist caves where often up to several thousand roost together.

Life Habits

This colonial bat feeds on insects. A single young is born in June.

Related Species

Relatives occur only in the tropical New World. Two occur in the southern part of this area: Mustached Bat (*Pteronotus parnellii),* chin not modified, wings attached to side of body, tail only half the length of interfemoral membrane; and Naked-backed Bat *(Pteronotus davyi),* wings uniquely attached to midline of back, giving naked-backed appearance. These have overlapping ranges in this area.

Ghost-faced Bat

Mustached Bat and Naked-backed Bat

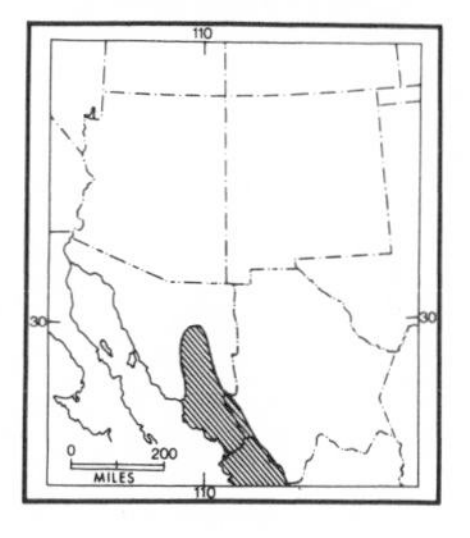

California Leaf-nosed Bat

Macrotus californicus

Order Chiroptera **Family Phyllostomatidae**

Identifying Features

This bat has a well developed wedge-shaped nose leaf (flap of flesh) on the tip of the nose. The ears, large and extending well beyond the tip of the snout, are joined at their bases. The long interfemoral membrane is naked.

Measurements

Total length, 3.8 inches (96 mm); tail, 1.3 inches (32 mm); hind foot, 0.6 inch (15 mm); ear, 1.3 inches (33 mm); forearm, 2 inches (50 mm); weight, 0.4 ounce (12 g).

Habitat

Restricted to low desert areas at elevations below 4,000 feet (1200 m). Day roosts are generally in caves and mine tunnels.

Life Habits

Colonial in habits, this bat is active throughout the year. Groups of up to several hundred spend the daytime in warm caves or mine tunnels and emerge at dark to feed. Being members of a primarily tropical family of bats, where low temperatures are rare, the Leaf-nosed Bats never evolved the ability to hibernate. On cold nights they remain in the warm cave or mine tunnel that serves as a day roost. Food is primarily night-flying insects, but in season ripe saguaro fruits are eaten. Fertilization occurs in September or October and after a long period of slow development, a single young is born the following June. The young grow rapidly, reaching adult size in six weeks.

Related Species

Relatives occur in tropical and subtropical New World. Only one, Waterhouse's Leaf-nosed Bat (*Macrotus waterhousii*), occurs in this region. It closely resembles the California Leaf-nosed Bat.

1. California
Leafed-nosed Bat

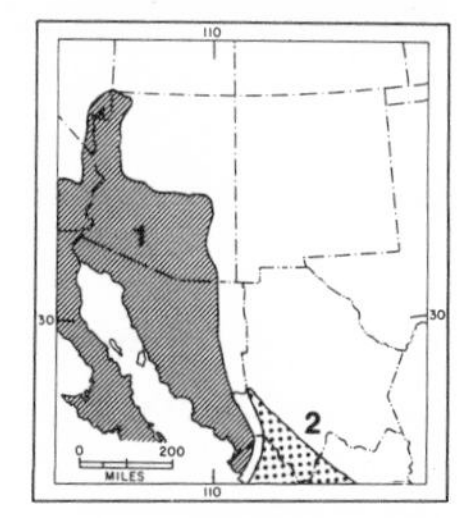

2. Waterhouse's
Leaf-nosed Bat

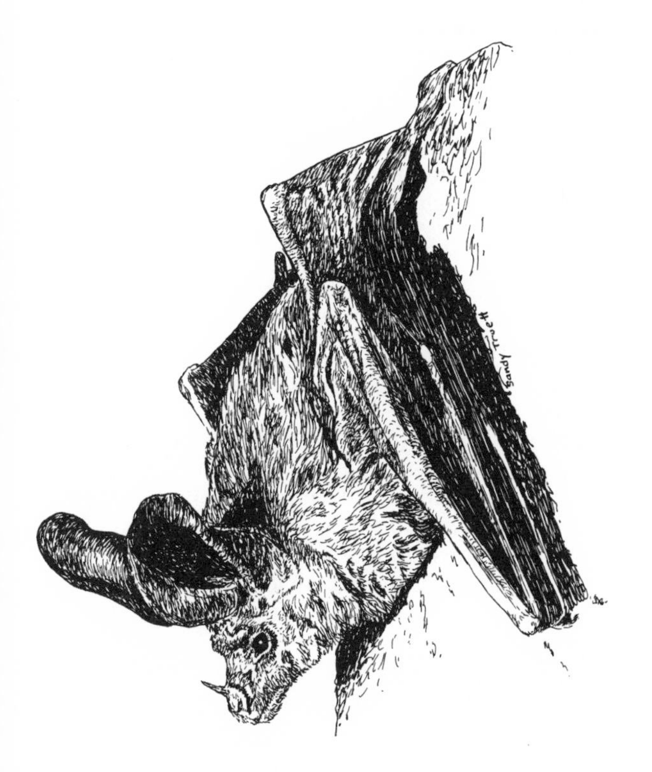

Long-tongued Bat

Choeronycteris mexicana

Order Chiroptera **Family Phyllostomatidae**

Identifying Features

These bats have the nose leaf (wedge-shape flap of flesh) characteristic of the family. Like Sanborn's Long-nosed Bat, they exhibit adaptations for feeding on nectar: the rostrum (snout) elongated, ears reduced, and tongue long and extensile. They have a short, naked, interfemoral membrane with an abbreviated tail that extends only about half the length of the interfemoral membrane.

Measurements

Total length, 3.5 inches (90 mm); tail, 1.6 inches (40 mm); hind foot, 0.5 inch (12 mm); ear, 0.7 inch (17 mm); forearm, 1.7 inches (43 mm); weight, 0.6 ounce (18 g).

Habitat

These bats roost, often singly but sometimes in small groups, in the twilight zone of shallow caves and entrances of mine tunnels.

Life Habits

This bat feeds on the nectar of various agaves and cacti. During late spring and summer females move northward into the mountain foothills of southern Arizona. During the winter they migrate to areas where nectar exists. In the northern part of the range a single young is born in June. Males apparently spend all summer in the more southern part of the range.

Related Species

Shrew-faced Bat (*Glossophaga soricina*), size smaller with forearm about 1.4 inches (75 mm), and the rostrum not so elongated.

1. Long-tongued Bat

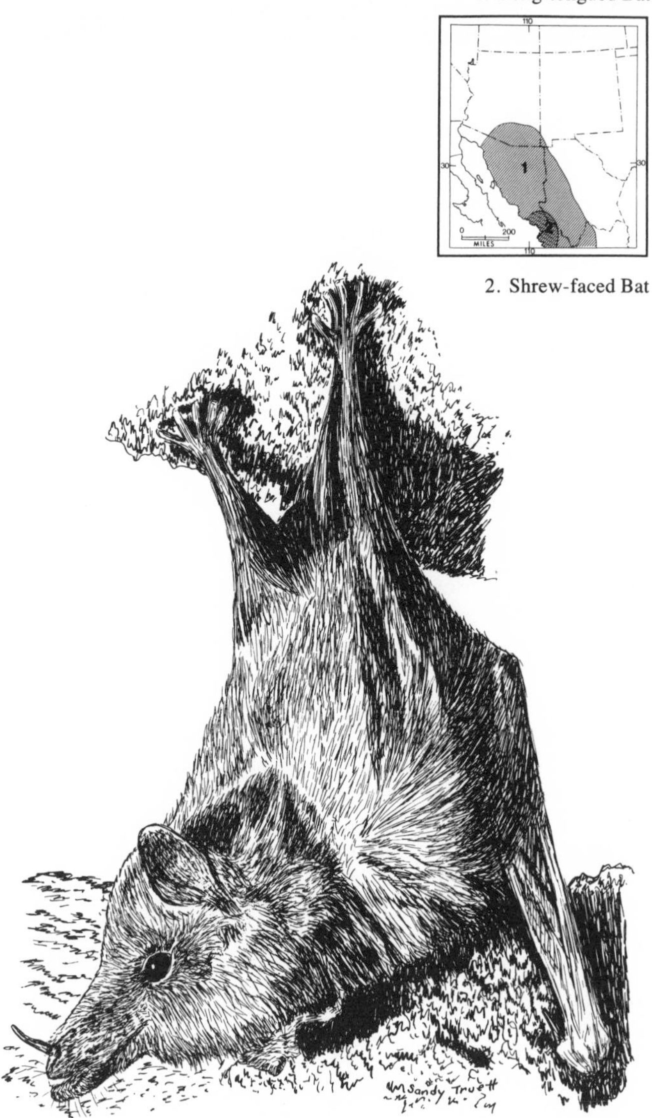

2. Shrew-faced Bat

Sanborn's Long-nosed Bat

Leptonycteris sanborni

Order Chiroptera **Family Phyllostomatidae**

Identifying Features

Similar to the Long-tongued Bat in shape and habits. The characteristic leaf nose is present, the rostrum (snout) is elongated and the ears are reduced. Differs especially in that the tail is absent and the interfemoral membrane is reduced to a narrow, fur-covered ridge.

Measurements

Total length, 3 inches (77 mm); tail, absent; hind foot, 0.6 inch (15 mm); ear, 0.6 inch (16 mm); forearm, 2.1 inches (53 mm); weight, 0.6 ounce (19 g).

Habitat

Day roosts are in caves and mine tunnels in the lower desert of the southern part of the area. Generally occurs near areas of native agaves and cacti at elevations below 5,000 feet (1500 m).

Life Habits

These bats feed on nectar of various agaves and cacti. In southern Arizona they seasonally feed extensively on saguaro nectar and serve as efficient pollinating agents. They also feed on the soft ripe fruit of the saguaros and probably that of other cacti. Females arrive at summer roosts in southern Arizona about the first of May. By mid-May each gives birth to a single young. By mid-October all are gone from the northern part of the range. During the fall they migrate to the more southern part of their range, with winter colonies being found from southern Sonora southward.

Related Species

The Mexican Long-nosed Bat (*Leptonycteris nivalis*) is slightly larger and has slightly longer wings.

1. Sanborn's
Long-nosed Bat

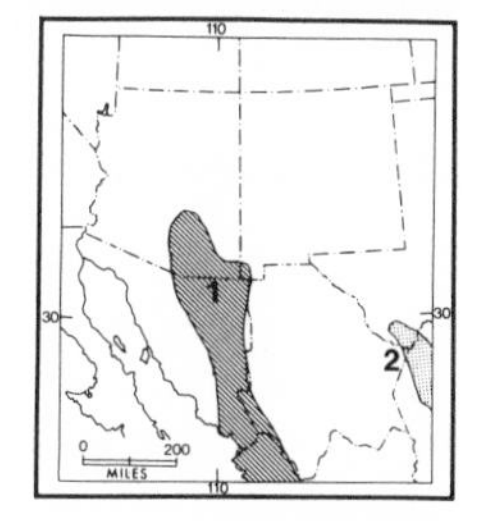

2. Mexican
Long-nosed Bat

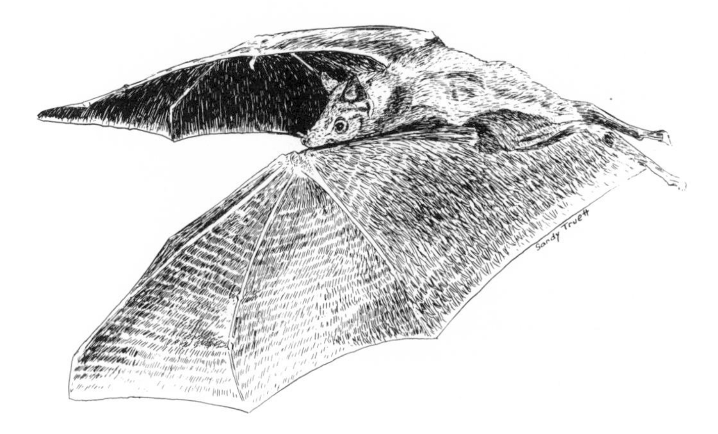

Little Brown Bat

Myotis lucifugus

Order Chiroptera **Family Vespertilionidae**

Identifying Features

This plain-nosed bat has no outgrowths of skin on the nose or lip. Its long tail is surrounded by the interfemoral membrane. The ears are short. The tragus (fleshy outgrowth on anterior part of ear) ends in a sharp point, in contrast to the rounded tip of the tragus in other species.

Measurements

Total length, 3.3 inches (85 mm); tail, 1.5 inches (38 mm); hind foot, 0.4 inch (9 mm); ear, 0.6 inch (14 mm); forearm, 1.5 inches (38 mm); weight, 0.4 ounce (10 g).

Habitat

Generally associated with forests. Summer roosts in attics of buildings, caves or mine tunnels. Hibernate in cold caves or mine tunnels.

Life Habits

During the warmer months these bats eat night-flying insects. Late in the warm season, they gain much weight from stored fat, enabling them to hibernate when no insects are flying. Winter roosts (hibernals) generally have uniform temperature throughout the season. Several thousand individuals may cluster together. During the spring, females congregate in warm roosts, forming maternity colonies where the single young is born. Banding studies reveal that some live as long as 23 years.

Related Species

Eight related species occur in the region: California Myotis (*Myotis californicus*); Cave Myotis (*Myotis velifer*); Fringed Myotis (*Myotis thysanodes*); Long-eared Myotis (*Myotis evotis*); Long-legged Myotis (*Myotis volans*); Small-footed Myotis (*Myotis leibii*); Southwestern Myotis (*Myotis auriculus*); and Yuma Myotis (*Myotis yumanensis*).

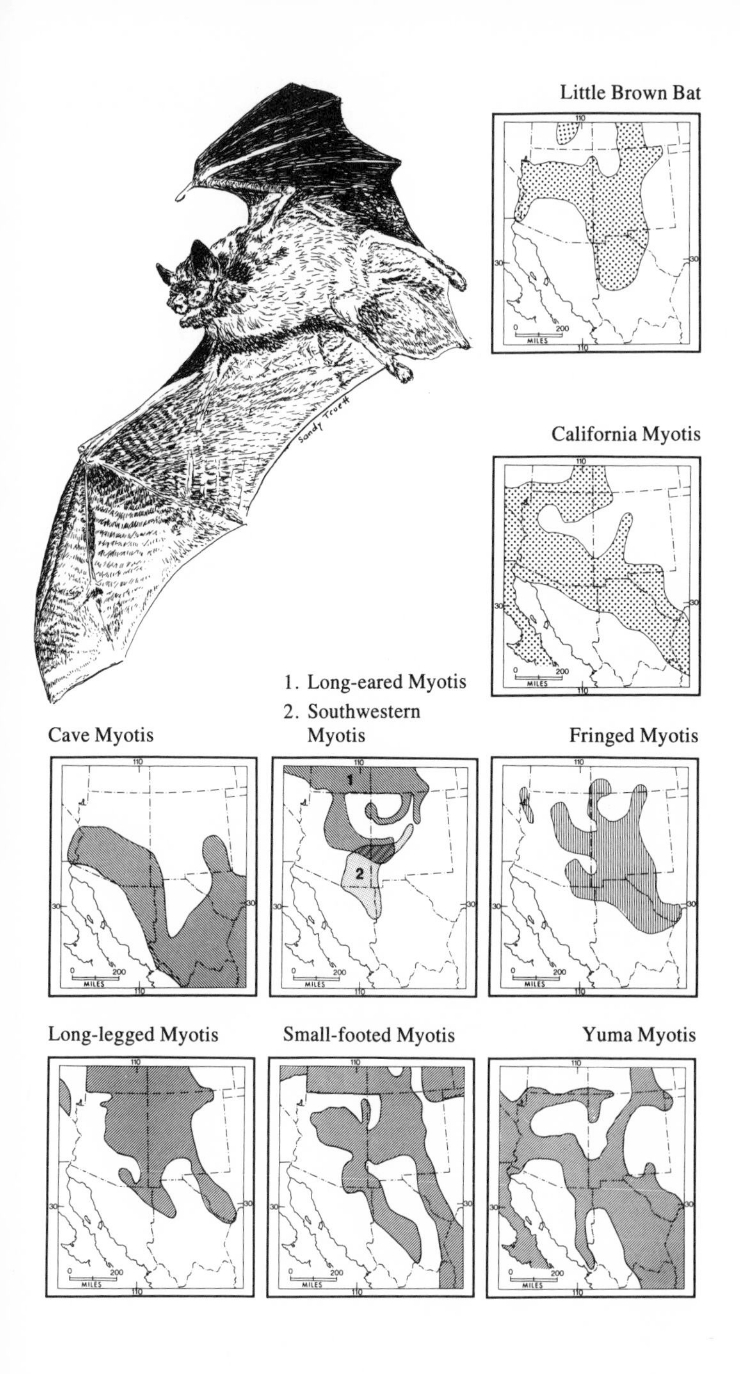

Sandy Truett
Little Brown Bat
California Myotis
1. Long-eared Myotis
2. Southwestern Myotis
Cave Myotis
Fringed Myotis
Long-legged Myotis
Small-footed Myotis
Yuma Myotis
110
30
0
200
MILES
1
2

Silver-haired Bat

Lasionycteris noctivagans

Order Chiroptera **Family Vespertilionidae**

Identifying Features

This plain-nosed bat has a long tail in a long interfemoral membrane. The basal half of the interfemoral membrane is covered with fur. The color is characteristic, almost black, with the tips of many hairs being white, resulting in the silver-haired appearance.

Measurements

Total length, 4 inches (102 mm); tail, 1.7 inches (43 mm); hind foot, 0.4 inch (9 mm); ear, 0.6 inch (16 mm); forearm, 1.6 inches (41 mm); weight, 0.2 ounce (7 g).

Habitat

In this area occurs in montane forests generally at elevations above 4,000 feet (1200 m).

Life Habits

These bats feed on small night-flying insects, mainly moths. Feeding flights generally begin after it is completely dark. Roosting is generally solitary in trees, often under bits of bark or in slight holes. One or two young are born in late June or early July. During cold months these bats generally migrate down mountains where on warm nights they feed, but on cold nights they remain inactive and hibernate. Some make a north-south seasonal migration, spending the summer months in southern Canada and the winter in southern United States.

Silver-haired Bat

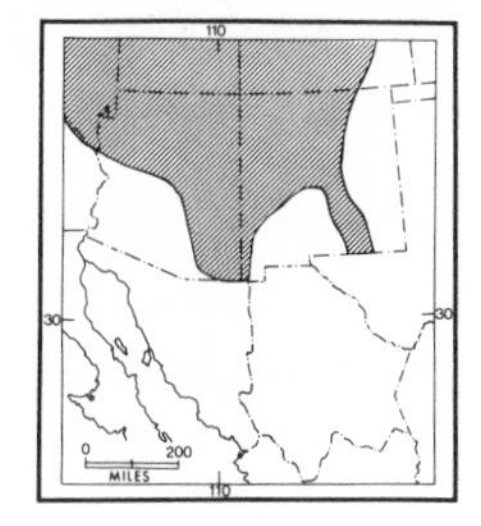

Western Pipistrelle

Pipistrellus hesperus

Order Chiroptera **Family Vespertilionidae**

Identifying Features

This plain-nosed bat is the smallest bat in the United States. The tail and interfemoral membrane are long. The ears have a short, rounded tragus (fleshy vertical growth on lower edge of ear). The color is light, buffy gray with a black mask across the eyes.

Measurements

Total length, 2.8 inches (72 mm); tail, 1.3 inches (32 mm); hind foot, 0.2 inch (5 mm); ear, 0.5 inch (13 mm); forearm, 1.2 inches (30 mm); weight, 0.1 ounce (4 g).

Habitat

Occurs most commonly in areas of rock exposures. Often extremely common in rock-walled canyons, usually at elevations below 5,000 feet (1500 m).

Life Habits

Pipistrelles are generally solitary, roosting in rock crevices in canyon walls. Small groups have been found roosting in crevices in brick buildings and even behind window shutters. The evening flight is early, often before sundown. These are the bats commonly seen at dusk feeding on small insects that are captured in flight. Two young per litter is common. The young are born in late June or early July. Males generally spend the winter months at lower elevations and actively feed on warm evenings, while females generally move to higher, colder elevations and hibernate.

Western Pipistrelle

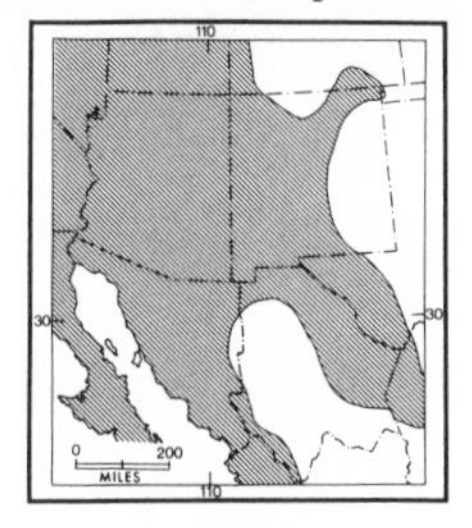

Big Brown Bat

Eptesicus fuscus

Order Chiroptera **Family Vespertilionidae**

Identifying Features

These plain-nosed bats have a long tail in a naked interfemoral membrane. The ears are short and the tragus (dorsal growth of flesh at base of ear) is blunt. Often confused with the Little Brown Bat and its related species. Big Brown Bats differ in having a blunt tragus and being larger in size. In general the wing and interfemoral membranes are almost black, much darker than those in the Little Brown Bat.

Measurements

Total length, 4.7 inches (119 mm); tail, 1.5 inches (39 mm); hind foot, 0.4 inch (11 mm); ear, 0.6 inch (16 mm); forearm, 1.9 inches (47 mm); weight, 0.7 ounce (20 g).

Habitat

Occurs at a wide range of elevations. Summer day roosts generally in buildings, hollow trees and similar situations. Winters are spent in hibernation in cold rock crevices and caves at higher elevations.

Life Habits

Food consists of night-flying insects. Summer day roosts are in a variety of warm situations where the female gives birth to one or two young, usually in mid-June. During the fall most migrate to cooler, higher elevations.

Big Brown Bat

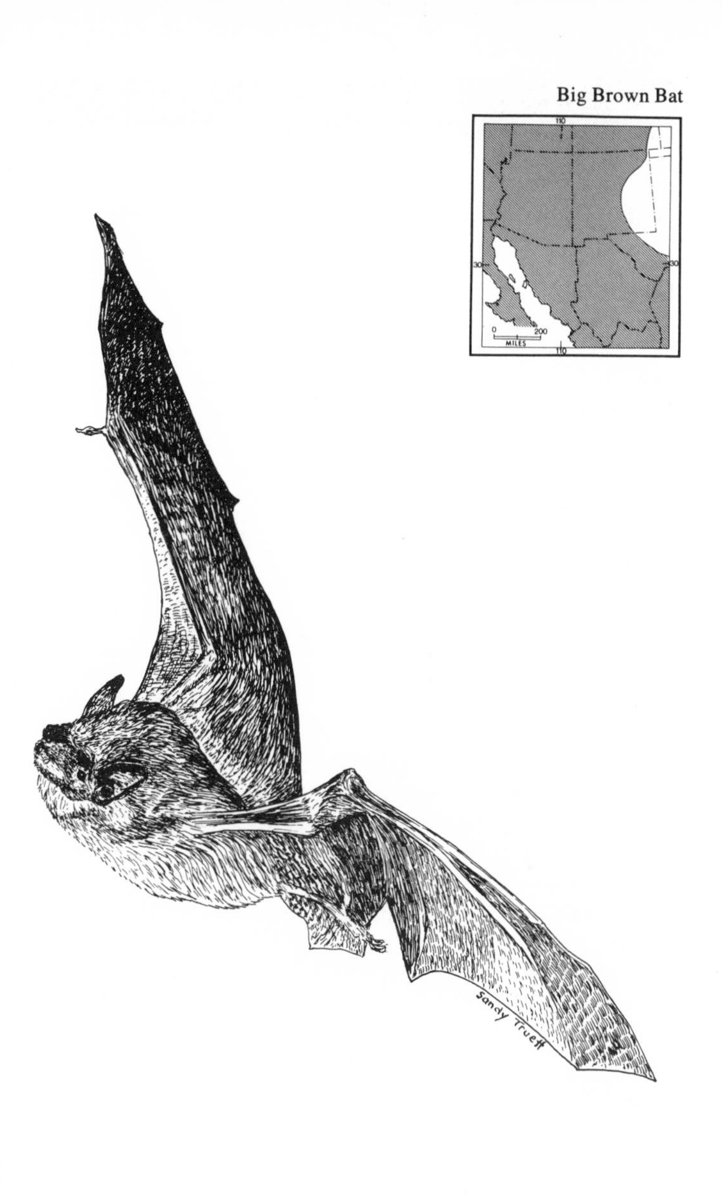

Red Bat

Lasiurus borealis

Order Chiroptera **Family Vespertilionidae**

Identifying Features

These plain-nosed bats have a long tail enclosed in a fur-covered interfemoral membrane. The ears are small. The pelage color is bright orange-red to buff, with scattered, individual white-tipped hairs. Size is medium.

Measurements

Total length, 4.3 inches (108 mm); tail, 1.7 inches (43 mm); hind foot, 0.3 inch (7 mm); ear, 0.4 inch (10 mm); forearm, 1.7 inches (42 mm); weight, 0.4 ounce (12 g).

Habitat

Tree roosting bats that occur in riparian habitats at lower elevations and in montane forests.

Life Habits

These bats feed on insects that are captured in flight after dark. Feeding generally occurs along the edges of tree clumps or over water surrounded by trees. Day roosts are in clumps of leaves in trees. Roosting is solitary except that a female hangs together with her young in a cluster. Two to four young are born in late June or early July. Seasonal migrations occur, with winters being spent at lower elevations in the south while summers are at high elevations or in the north.

Red Bat

Hoary Bat

Lasiurus cinereus

Order Chiroptera **Family Vespertilionidae**

Identifying Features

Similar to the Red Bat but much larger and with dorsal color of yellowish to dark brown and a "hoary" appearance resulting from many hairs being white-tipped.

Measurements

Total length, 5.6 inches (142 mm); tail, 1.9 inches (49 mm); hind foot, 0.5 inch (13 mm); ear, 0.7 inch (17 mm); forearm, 2.1 inches (53 mm); weight, 1 ounce (28 g).

Habitat

A tree bat, restricted to areas where trees are present.

Life Habits

Food made up entirely of night-flying insects. During the day this bat hangs solitarily in leaves of a tree. Females give birth to two young, generally in late June or early July. Seasonal migrations involve movements from northern areas southward or from high in southern mountains downslope to low elevations. Hoary Bats are strong fliers and probably make relatively long flights.

Hoary Bat

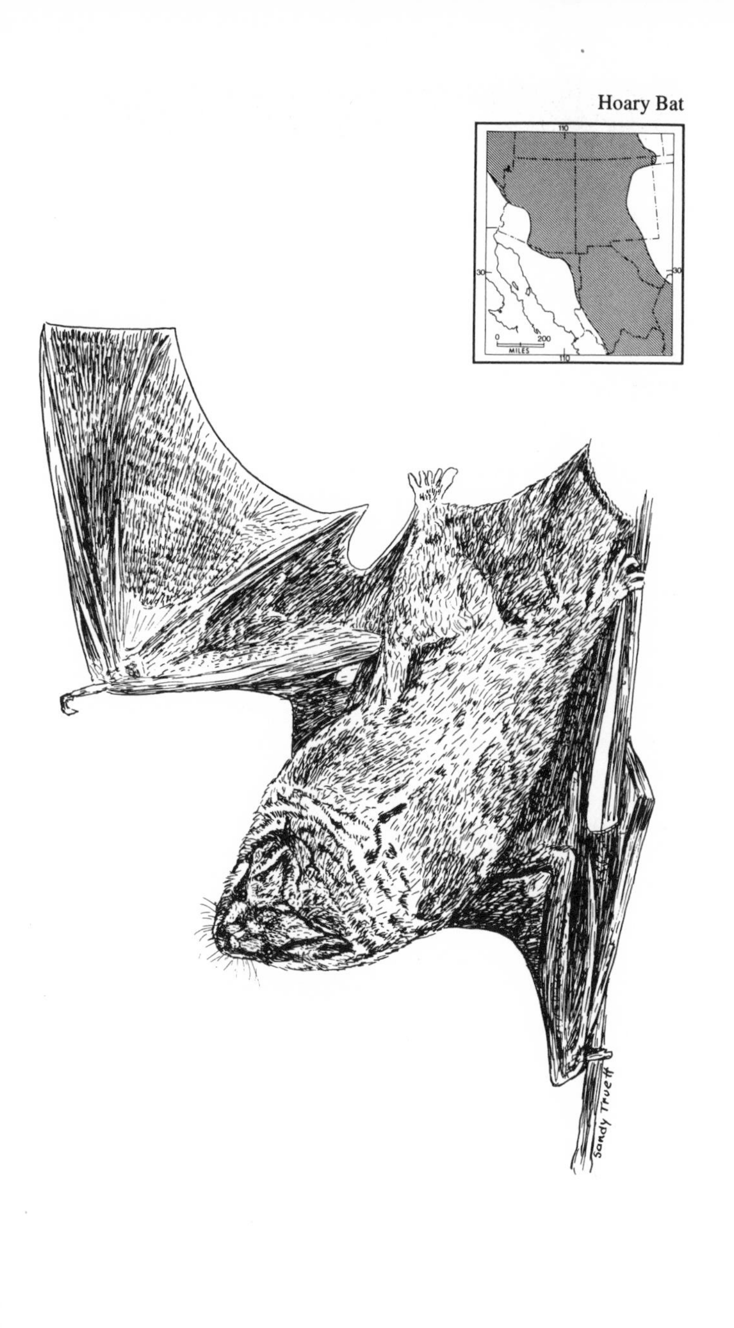

Southern Yellow Bat

Lasiurus ega

Order Chiroptera **Family Vespertilionidae**

Identifying Features

Similar to the Red Bat but slightly larger and with only the basal half of the interfemoral membrane covered by hair. Color is yellowish to buffy, with only a few hairs tipped with white.

Measurements

Total length, 4.4 inches (112 mm); tail, 2 inches (51 mm); hind foot, 0.4 inch (9 mm); ear, 0.6 inch (16 mm); forearm, 1.9 inches (49 mm); weight, 0.5 ounce (13 g).

Habitat

Primarily associated with palm trees.

Life Habits

Food made up entirely of night-flying insects. A solitary bat with day roosts usually near the bases of leaf fronds of palm trees. Two young per litter born in early June is apparently normal. This is essentially a tropical species that occurs only in the southern part of this area.

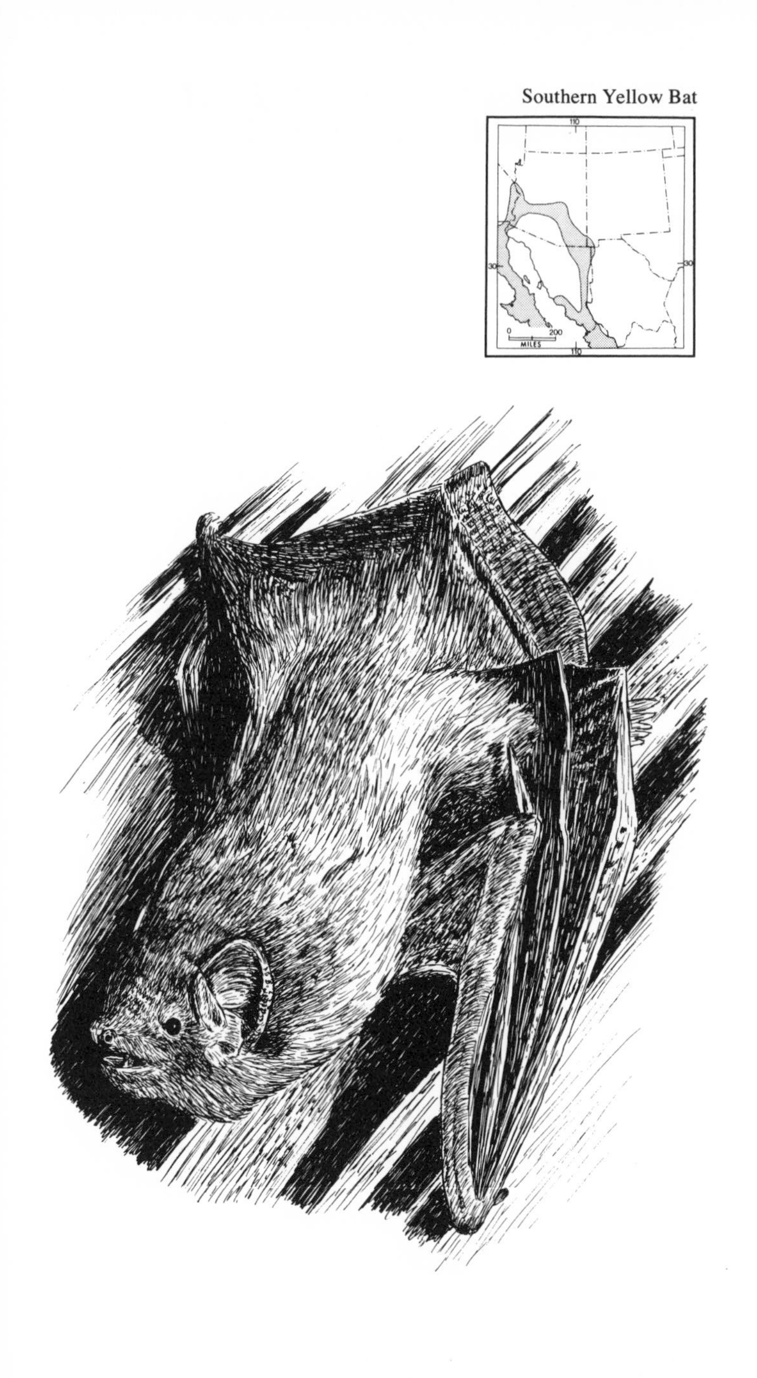

Southern Yellow Bat

Spotted Bat

Euderma maculatum

Order Chiroptera **Family Vespertilionidae**

Identifying Features

This plain-nosed bat has unique, extremely large, pink-colored ears and a black pelage with three dorsal white spots, one on each shoulder and one on the rump. The long tail is enclosed in a naked interfemoral membrane.

Measurements

Total length, 4.3 inches (110 mm); tail, 1.9 inches (48 mm); hind foot, 0.4 inch (11 mm); ear, 1.8 inches (45 mm); forearm, 1.9 inches (49 mm); weight, 0.6 ounce (17 g).

Habitat

A rarely seen bat whose requirements are not fully understood. Has been taken from hot, low desert to elevations of 8,000 feet (2400 m) in mountains.

Life Habits

Food consists of night-flying insects. Day roosts are not known but probably are in deep rock crevices in high canyon walls. Currently it is known from scattered localities in the western United States and from a single locality in Chihuahua.

Spotted Bat

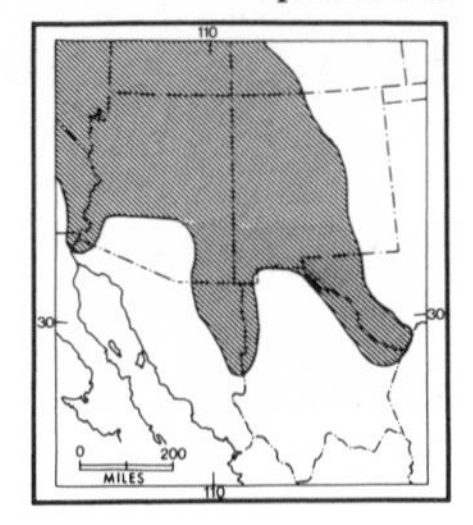

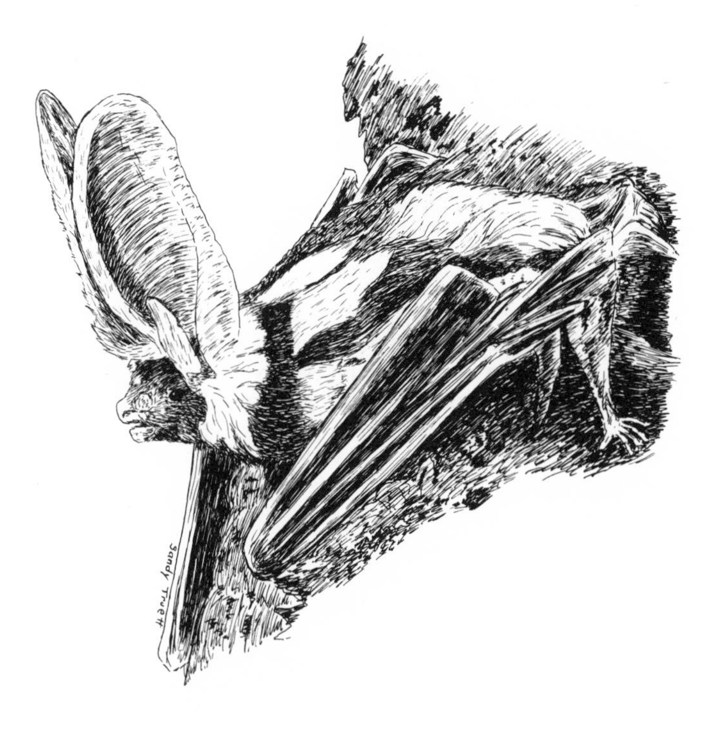

Allen's Big-eared Bat

Idionycteris phyllotis

Order Chiroptera **Family Vespertilionidae**

Identifying Features

This large-eared, plain-nosed bat has a long tail encased in a long, naked interfemoral membrane. It is distinguished from other American bats by the presence of two prominent flaps of skin between the two large ears.

Measurements

Total length, 4.6 inches (117 mm); tail, 2.2 inches (55 mm); hind foot, 0.4 inch (10 mm); ear, 1.6 inches (41 mm); forearm, 1.9 inches (47 mm); weight, 0.4 ounce (11 g).

Habitat

Occurs in wooded and forested regions, generally above 4,000 feet (1200 m). Day roosts generally in rock crevices.

Life Habits

These bats feed entirely on night-flying insects. During the summer small numbers (up to 50) congregate in a maternity colony. There the single young is born, usually in mid-June. Winter roosts are not yet known but probably are in rock crevices at higher elevations in the mountains.

Related Species

The Townsend's Big-eared Bat (*Plecotus townsendii*) is slightly smaller (forearm about 1.7 inches or 47 mm), lacks the basal ear flaps and has large glandular lumps on each side of the rostrum. The Mexican Big-eared Bat (*Plecotus mexicanus*) is known from part of the Sierra Nevada range.

Allen's Big-eared Bat

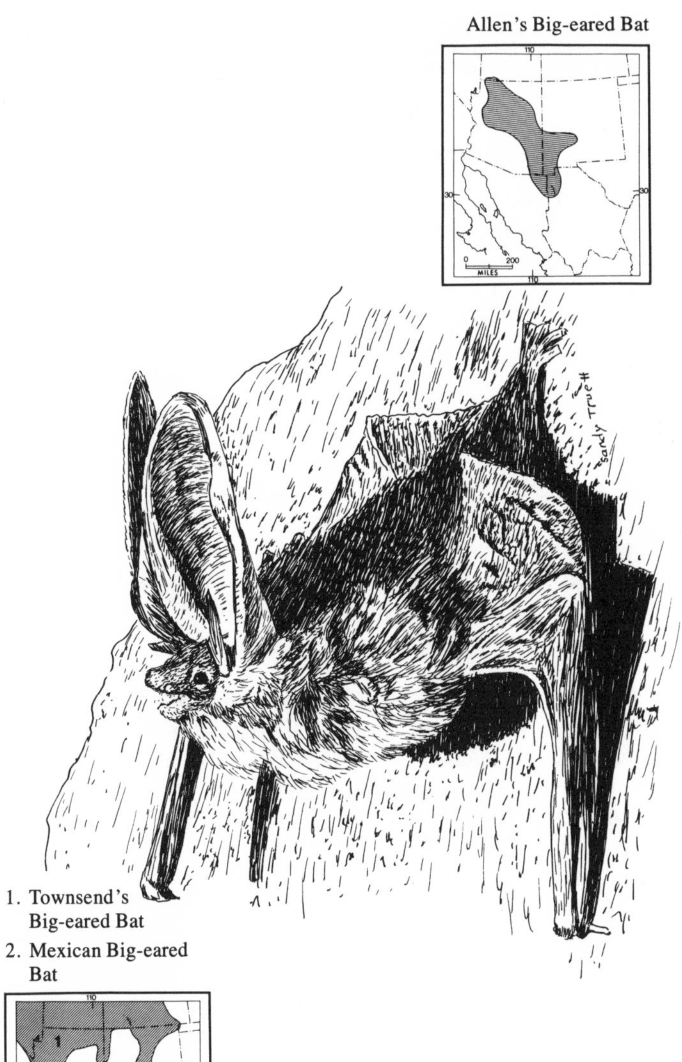

1. Townsend's Big-eared Bat
2. Mexican Big-eared Bat

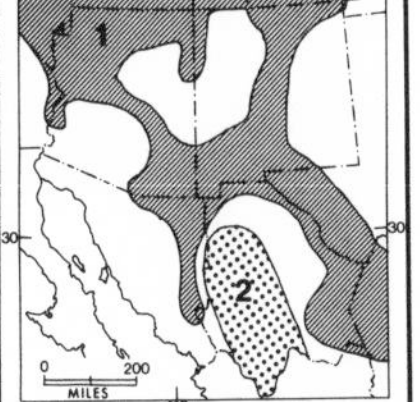

Pallid Bat

Antrozous pallidus

Order Chiroptera **Family Vespertilionidae**

Identifying Features

This plain-nosed, big-eared bat has a long tail encased in a long interfemoral membrane. The body structure is heavy for a bat, with strong wing bones. The color is light with individual hairs creamy white, tipped with brown or black. The ears are large.

Measurements

Total length, 4.7 inches (120 mm); tail, 1.8 inches (46 mm); hind foot, 0.4 inch (10 mm); ear, 1.2 inches (30 mm); forearm, 2.2 inches (55 mm); weight, 0.7 ounce (20 g).

Habitat

Generally below 4,500 feet (1350 m), usually in desert grasslands or areas of rocky outcrops in the desert.

Life Habits

These bats feed on a variety of large arthropods, often captured on the surface of the ground. Included are sphinx moths, scarab beetles, grasshoppers, crickets, and scorpions. The food item is carried to a night roost in shallow caves, mine tunnels and even in carports, porches and patios of houses, where the soft parts are eaten while the hard parts, such as wings and legs, are dropped to the ground. Day roosts are in attics of buildings, in crevices under bridges, rock crevices and similar situations. Two (sometimes one) young are born in June. Winter is spent in hibernation, usually in rock crevices at colder, higher elevations.

Pallid Bat

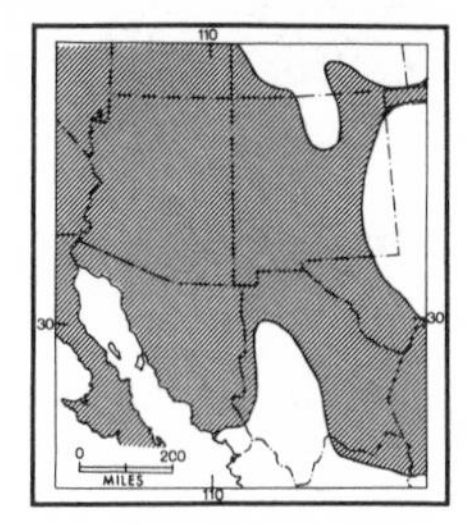

Brazilian Free-tailed Bat

Tadarida brasiliensis

Order Chiroptera **Family Molossidae**

Identifying Features

Free-tailed bats have a significant portion of the tail extending posterior to the interfemoral membrane. The flight membranes are dark, thick and leathery. The ears are flattened, thickened and extend forward over the eyes. The hind feet have well developed tactile hairs.

Measurements

Total length, 4 inches (102 mm); tail, 1.3 inches (34 mm); hind foot, 0.4 inch (10 mm); ear, 0.7 inch (17 mm); forearm, 1.7 inches (42 mm); weight, 0.4 ounce (12 g).

Habitat

Most common at lower elevations in the southern part of the area but occur, at least during the summer, from high mountains to low deserts.

Life Habits

A colonial bat, often congregating in colonies of millions such as at Carlsbad Caverns. They feed on night-flying insects, especially small moths. Generally, they overtake a flying moth from the rear, bite off the soft abdomen and let the wings, legs and thorax fall to the ground. Large maternity colonies are formed in the late spring where the single young is born, usually in late June. During the winter most of these bats migrate southward into Mexico.

Related Species

In the area are two similar, related species. Both are larger: Pocketed Free-tailed Bat (*Tadarida femorosacca*), forearm, 1.9 inches (48 mm); weight 0.5 ounce (14 g). Big Free-tailed Bat (*Tadarida macrotis*) forearm, 2.4 inches (60 mm); weight, 0.9 ounce (26 g).

Brazilian Free-tailed Bat

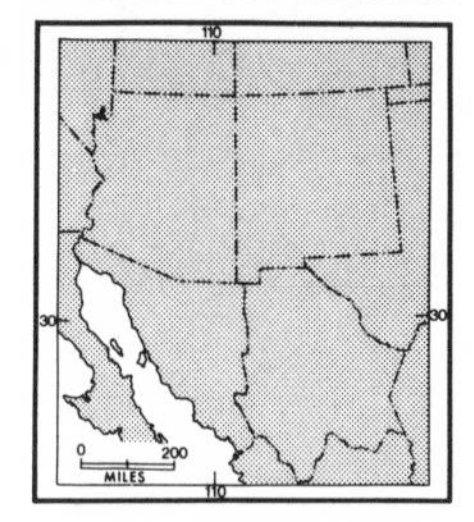

and Big Free-tailed Bat

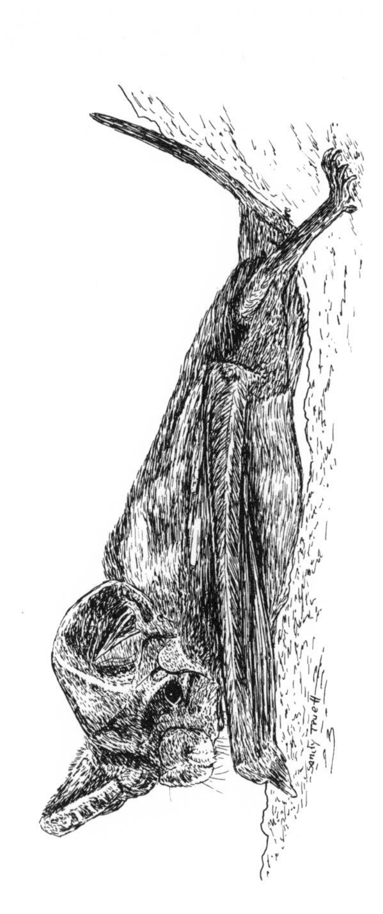

Pocketed Free-tailed Bat

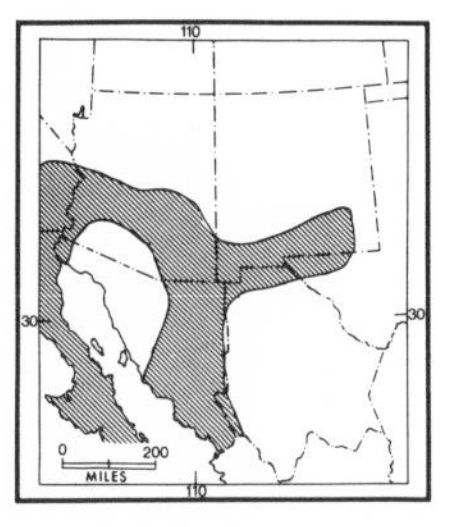

Western Mastiff Bat

Eumops perotis

Order Chiroptera **Family Molossidae**

Identifying Features

Like the Brazilian Free-tailed Bat, this species has a free tail, thick flight membranes, and tactile hairs on the feet. They differ in being much larger. The low flattened ears are widely fused together at the inner base. The largest bat in the United States, it has a wingspread of 21 inches (530 mm).

Measurements

Total length, 2.7 inches (184 mm); tail 2.2 inches (55 mm); hind foot, 0.7 inch (17 mm); ear, 1.6 inches (41 mm); forearm, 3.0 inches (76 mm); weight, 2.1 ounces (60 g).

Habitat

Generally in areas with high rocky cliffs near larger bodies of open water at the edges of the desert.

Life Habits

These bats feed entirely on night-flying insects. Day roosts are in small colonies (50–60 individuals) in rock crevices high above a canyon floor. A few roosts have been found in the attics of two-story buildings. Having very narrow wings, these bats have difficulty taking off from a flat surface and must climb something. They take off by dropping down, setting their wings and then becoming airborne. The single young is born in late June or early July.

Related Species

A single, slightly smaller related species occurs in this area: Underwood's Mastiff Bat (*Eumops underwoodi*), forearm, 2.7 inches (68 mm); weight 1.9 ounces (55 g).

1. Western Mastiff Bat

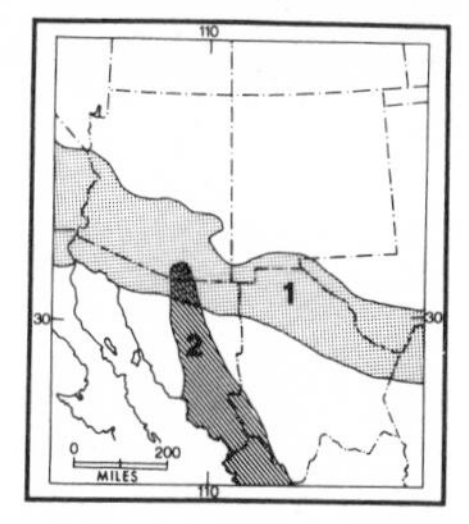

2. Underwood's Mastiff Bat

Vagrant Shrew

Sorex vagrans Baird

Order Insectivora **Family Soricidae**

Identifying Features

A small mammal with a long snout and short, dense, velvetlike black fur. The eyes and ears are so reduced that they are difficult to see. The teeth are numerous, and most have red pigment in the enamel.

Measurements

Total length, 4 inches (104 mm); tail, 1.6 inches (41 mm); hind foot, 0.5 inch (12 mm); weight, 0.3 ounce (8 g).

Habitat

Montane animals that usually live in ponderosa pines at higher elevations. Most common along streams where humus is moist and thick and plant cover is dense.

Life Habits

Shrews are active night and day. They require much food, often eating up to three quarters of their total body weight of food in a day. Insects make up most of the food, but earthworms and even small mice are fed upon. Most of their lives are spent in a small area, probably less than a small city lot. Usually one litter of young is born each summer after a gestation period of about 20 days. Two to nine (usually four to six) young are born in each litter.

Related Species

Six other kinds of shrews occur in the region: Arizona Shrew (*Sorex arizonae*), smaller than Vagrant Shrew; Merriam's Shrew *(Sorex merriami),* similar to the Arizona Shrew; Dwarf Shrew (*Sorex nanus*), restricted to bogs; Masked Shrew (*Sorex cinereus*), inhabits mountains above 9,000 feet (2700 m); Water Shrew (*Sorex palustris*), largest, lives near permanent mountain streams; and Desert Shrew (*Notiosorex crawfordi*), ears evident above surface of the fur.

Vagrant Shrew

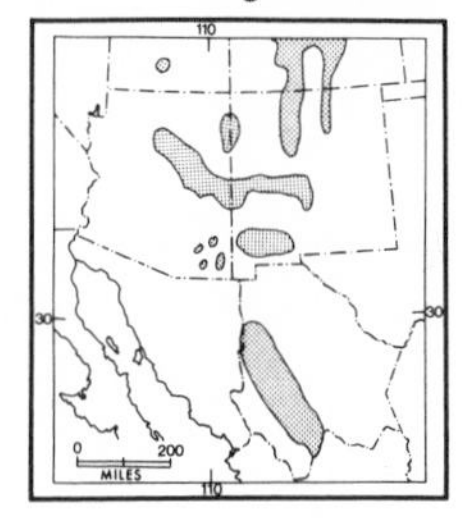

1. Desert Shrew
2. Water Shrew

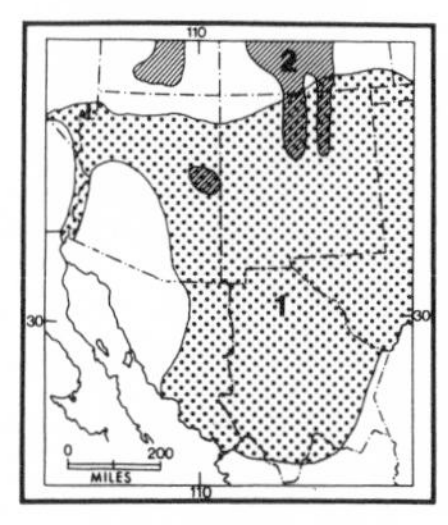

1. Merriam's Shrew
2. Arizona Shrew

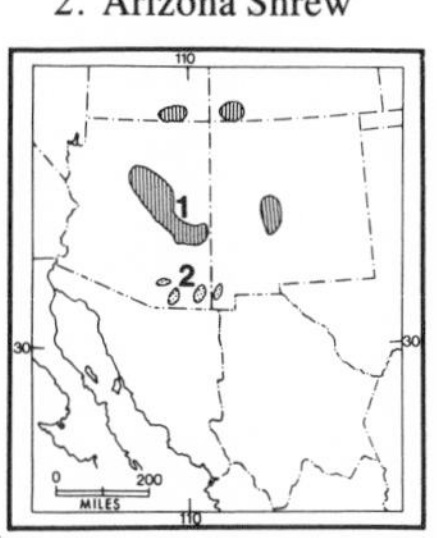

1. Masked Shrew
2. Dwarf Shrew

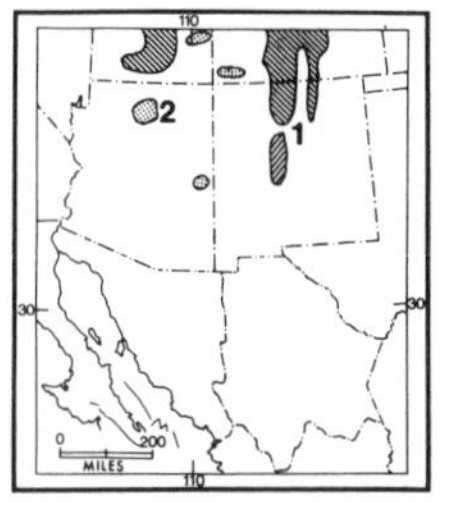

Suggested Readings

Anderson, S. 1972. *Mammals of Chihuahua, Taxonomy and Distribution*. Bull. Amer. Museum Nat. Hist., 148(2):1–410.

Armstrong, D. M. 1972. *Distribution of Mammals in Colorado*. Univ. of Kansas, Museum of Natural History Monograph, 3:x+415.

Burt, W. H. and R. P. Grossenheider. 1976. *A Field Guide to the Mammals*. Houghton Mifflin (third edition) pp. xxvii+289.

Cockrum, E. L. 1960. *The Recent Mammals of Arizona: Their Taxonomy and Distribution*. The University of Arizona Press, pp. viii+276.

Editors. 1981. *Book of Mammals*. vol. one. A–J, pp. 1–304, vol. two. K–Z, pp. 305–608, National Geographic Society.

Editors. 1979. *Wild Animals of North America*. National Geographic Society, pp. 1–406.

Findley, J. S., A. H. Harris, D. E. Wilson, and C. Jones. 1975. *Mammals of New Mexico*. University of New Mexico Press, xxii+360.

Hoffmeister, D. F. and F. E. Durham. 1971. *Mammals of the Arizona Strip Including the Grand Canyon National Monument*. Museum of Northern Arizona, Tech. Services, 11:1–44.

Murie, O. J. 1974. *A Field Guide to Animal Tracks*. Houghton Mifflin (second edition) pp. xxi+375.

Olin, G. 1961. *Mammals of the Southwest Mountains and Mesas*. Globe, Arizona: Southwestern Monuments Association, pp. xv+126.

Olin, G. 1965. *Mammals of the Southwest Deserts*. Globe, Arizona: Southwest Monuments Association (third edition) pp. 1–112.

Whitaker, J. D., Jr. 1980. *The Audubon Society Field Guide to North American Mammals*. Alfred A. Knopf, pp. 1–745.

Index

Species in regular type are the major species in the book. Those in italics are related species.